20 70696029 0406

Surveying for Archæologists

Second edition

by

F. Bettess

M.Eng. (Liverpool)
Fellow of the Institution of Civil Engineers
Former Head of Civil Engineering, Sunderland Polytechnic

Durham
1992

First published 1984
Revised edition published 1992 by

Department of Archæology
University of Durham

Graphics by Yvonne Brown

Edited by A. F. Harding

Cover photograph taken by P. Bettess. at the Arbeia roman site, South Shields, by kind permission of Arbeia roman site.

Layout design, type-setting and production by Penshaw Press. Cleadon

Copyright ©1992 by Department of Archæology, University of Durham

BETTESS, F
    Surveying for Archæologists – 2Rev.ed
I. Title
526.902493
ISBN 0-905096-09-6

All rights reserved

No part of this publication may be reproduced, stored in a retrieval system, or transmitted in any form or by any means, electronic, mechanical, photocopying, recording or otherwise, without the prior written permission of the publishers.

Printed and bound in Great Britain by
Dotesios Limited,
Kennet House, Kennet Way,
Trowbridge, Wiltshire, BA14 8RN

Foreword

This book developed from my experience of being responsible for the surveying on a number of archæological excavations. It includes methods and techniques which I have found useful when surveying archæological sites, either as a preliminary to an excavation or for record purposes.

Fieldwork forms an important part of other disciplines and as in archæology surveying has a supporting rôle to play in providing the framework and control for recording purposes. I hope that workers in such areas will find this book useful.

Most of the methods described are fundamental to all practical surveying. In the early stages the object should be to master the techniques involved. Good technique is a combination of knowing and doing. It will only be acquired by practice. It involves the proper co-ordination of muscles, senses, judgement and intellect. You need to think about what you are doing, how you are doing it and why you are doing it that way.

I knew nothing of archæology until my wife led me into it as a hobby. My earlier archæological tutors were Miss Barbara Harbottle and Professor Rosemary Cramp. I am very grateful to them for making my introduction to archæology both interesting and pleasant. In addition they had sufficient confidence in my surveying to let me assume responsibility for it on their sites. My wife and I have made many friends because of our archæological activities but two of these deserve mention. Dr David Breeze and Professor Chris Morris were kind enough to give us the opportunity to widen our archæological surveying both in scale and geographical extent. The preparation of even a modest booklet like this requires a guide, counsellor and friend. Professor Anthony Harding has fulfilled this rôle so successfully that he is still a friend.

I owe a special debt of gratitude to my wife, Gladys. First of all for the shared interest of a lifetime. Then within that interest she has been an enthusiastic supporter of my surveying activities. She has acted as chainman, staffman and all the other supporting rôles. Together we have developed many of the techniques described, refining them from earlier trials. Wherever the word *I* appears, it should really be read as *We*.

Alnmouth, November 1983   F. Bettess

Contents

1 **Introduction to Archæological Surveying** 1
   1.1   Introduction 1
   1.2   Accuracy 3
   1.3   Scale 4

2 **Rectangular Grids** 6
   2.1   The Site and Grid 6
   2.2   Choice of Origin and Axes 7
   2.3   Setting out the Grid 8
   2.4   Setting out a Right-Angle 10
   2.5   Uneven Ground 14
   2.6   Chains 15
   2.7   Reference Points 16
   2.8   False Origin 16

3 **Features and Topography** 18
   3.1   Frameworks 18
   3.2   Linear Triangulation 19
   3.3   Angular Triangulation 20
   3.4   Traverses 21
   3.5   Relating Detail to Framework 21
   3.6   Optical Square 26
   3.7   Cross-Staff 27
   3.8   Prismatic Compass 27
   3.9   Connecting to Ordnance Survey 28
   3.10 Detail 28
   3.11 Plotting 29
   3.12 Drawings 30

4 **The Plane Table** 31
   4.1   Equipment 31
   4.2   Outline of Intersection Method 32
   4.3   Intersection Method Procedure 33
   4.4   Development of Method 35
   4.5   Radial Line Plotting 36

5 **Levelling** 38
   5.1   Basic Principles 38
   5.2   Types of Levels 39
   5.3   Setting Up - Dumpy 40
   5.4   Setting Up - Quickset 41
   5.5   Automatic Level 42
   5.6   Datum 42
   5.7   Bench-Marks 43
   5.8   Basic Levelling 43
   5.9   Booking and Reduction 45
   5.10 Setting Pegs to a Level 48
   5.11 Section Datum 49
   5.12 Abney Level 49

	5.13	Clinometer	50
	5.14	Boning Rods	50
6	**Earthworks**		52
	6.1	Plotting Conventions	52
	6.2	Hachures	52
	6.3	Contours	53
	6.4	Types of Contouring	55
	6.5	Contour Chasing	55
	6.6	Contour Interpolation - Grids	56
	6.7	Contour Interpolation - Spot Heights	56
	6.8	Assistants	56
	6.9	Close Contouring	58
7	**Theodolites - Basic Usage**		60
	7.1	Types of Theodolite	60
	7.2	Handling	60
	7.3	Plumb-bob, Optical Plummet	61
	7.4	Plumbing Rod	62
	7.5	Setting Up - Centring	62
	7.6	Setting Up - Levelling	63
	7.7	Setting Horizontal Angles	63
	7.8	Reading Angle Scales	64
	7.9	Vertical Angles	66
8	**Tacheometry**		67
	8.1	Basis of Method	67
	8.2	Inclined Sights	68
	8.3	Formulæ	69
	8.4	Application of method	71
	8.5	Station Routine	72
	8.6	Plotting	73
	8.7	Computer Processing	74
9	**Surveying Calculations**		76
	9.1	Introduction	76
	9.2	Sine, Cos and Tan	77
	9.3	Large Angles and Bearings	80
	9.4	The Sine Rule and Triangulation	86
	9.5	Plotting Triangulation	89
	9.6	Dealing with Sloping Ground	94
	9.7	Setting up a Grid on a Slope	96
	9.8	Problems with Large Grids	99
	9.9	Surveying of Detail	108
	9.10	Traverse Calculations	110
10	**EDM, Total Stations and Computers**		113
	10.1	Introduction	113
	10.2	Electronic Distance Measurement ...	113
	10.3	Total Station	114
	10.4	Setting Up	115
	10.5	Taking Readings	115
	10.6	Accuracy of EDM	118

10.7	Basic Co-ordinates	118
10.8	Applications of EDM and Total Stations	121
10.9	Picking Up Detail	121
10.10	Setting Out	123
10.11	Linear and Angular Measurement of Frameworks	124
10.12	Frameworks	125
10.13	Communication	127
10.14	Computers	128
10.15	Computer Software	128
10.16	Conclusion	130

Equipment check list 131

Further reading 132

Subject Index 134

LIST OF FIGURES

2.1	A grid superimposed on a plan	6
2.2	Locating grid on plan	7
2.3	Ranging a pole onto a straight line	9
2.4	Ranging poles, as seen by sighter	9
2.5	Setting out a right-angle using tapes	11
2.6	Setting out a right-angle using an integer triangle	12
2.7	Right angle triangle with two equal sides	12
2.8	Extending the grid	14
2.9	Measuring distances on a slope	14
2.10	Location of a reference point	16
2.11	Using a false origin for grid references	17
3.1	Possible errors from assumption of right-angles	18
3.2	Framework around building	18
3.3	Use of check measurements	19
3.4	Intersection of arcs to locate new point	19
3.5	Good and poor intersections	19
3.6	Open and closed traverses	21
3.7	Relating detail (point P) to framework (line AB)	22
3.8	Use of 'minimum distance' for obtaining perpendicularity	22
3.9	Locating points on a feature from a tape line	23
3.10	Locating features in a trench	24
3.11	Transferring the framework to the bottom of a trench	25
3.12	Grid in trench	26
4.1	The plumbing fork	31
4.2	Site with baseline	33
4.3	Plane table at A	34
4.4	Plane table at B	35
4.5	Zone of good intersections on plane table	36
4.6	Use of a plane table traverse	36
5.1	Level plane defined by a bowl full of water	38
5.2	Use of a spirit bubble to define level plane	39
5.3	Three screw levelling with bubble	40

5.4	Ordnance Survey bench-mark	43
5.5	Transferring the level	45
5.6	Use of boning rods	51
6.1	Various styles of hachuring	52
6.2	Contour ambiguities	53
6.3	Contours for a complex earthwork	54
6.4	Contour chasing	55
6.5	Rectangular and triangular grids	56
6.6	Contour interpolation from spot heights	56
8.1	Typical stadia lines	67
8.2	Using level as tacheometer	67
8.3	Using theodolite as a tacheometer	69
8.4	Finding level by tacheometer	70
8.5	Triangulation system used at Barhill Roman Fort	71
8.6	Measured and horizontal lengths	72
8.7	Tacheometric survey readings	73
9.1	Right angled triangle, showing $\sin A$	77
9.2	Right angled triangle, scaling effect	77
9.3	Right angled triangle, showing $\sin A$, $\cos A$ and $\tan A$	78
9.4	Limiting case for $A \to 0°$ and $A \to 90°$	79
9.5	General case of right angled triangle	80
9.6	Mathematical bearing convention - from x direction	80
9.7	Surveying bearing convention - from North	80
9.8	Sin of angle between 90° and 180°	81
9.9	Sign convention, latitude and departure	81
9.10	Quadrant bearings	83
9.11	Example of 30° bearing	83
9.12	Example of 150° bearing	84
9.13	Example of 210° bearing	85
9.14	Sine does not change with the type of triangle	86
9.15	The sine rule	86
9.16	Simple triangulated quadrilateral	87
9.17	Small triangulated network	88
9.18	Sequence of side calculation	89
9.19	Two different bearings of the same line	90
9.20	Bearings in a small triangulation	90
9.21	Bearings, latitudes and departures	91
9.22	Co-ordinates of triangulation points	92
9.23	Continuing a line down a slope	94
9.24	Setting out a point down a slope	95
9.25	Dip and strike	96

9.26	Base line with lines at right-angles	97
9.27	Setting points on one line	98
9.28	Co-ordinates of point in grid	99
9.29	Co-ordinates of three points in grid	100
9.30	Setting out a corner of a large grid	104
9.31	Setting out grid points from grid points	105
9.32	Triangulation station near area of interest	106
9.33	Main grid pegs	106
9.34	Triangle with subsidiary point	106
9.35	Angular measurement from two known points	108
9.36	Traverse calculations	110
10.1	Theodolite, theodolite with EDM, and total station	114
10.2	Simple EDM controls	115
10.3	Total station key pad	116
10.4	Vertical elevation of EDM and target	119
10.5	Co-ordinate relationships	120
10.6	Sketch plan	122
10.7	Setting out a grid	123
10.8	Triangulation using polygon with central station	125
10.9	Trilateration using polygon with central station	126

LIST OF TABLES

1.1	Comparison of Scales	4
2.1	Diagonal values, D, for triangles with two sides of length S	12
5.1	Booking of Levels	45
5.2	Booking and Reduction of Levels	47
9.1	Sine, cosine and tangent signs in the four quadrants	83

Chapter 1

Introduction to Archæological Surveying

1.1 Introduction

Archæological surveying is a skill which many archæology students and excavators believe, wrongly, to be beyond their comprehension. Although the basic techniques are extremely straightforward, many people, even those with plenty of experience in the field, frequently find inordinate difficulties in getting to grips with them. Yet a clear appraisal of the skills involved, and a little practice in the field, can turn the beginner into a valuable surveying assistant, or more, very quickly. Indeed, without such basic surveying skills, no field archæologist should be venturing further into the serious business of site destruction known as excavation. It is in recognition of that fact that most British university courses in archæology now include basic surveying as a compulsory, though not always examined, part of the degree course. This booklet aims to help archæology students and other fieldworkers to achieve a respectable level of surveying skill with a minimum of technical expertise.

Usually the archæologist first meets surveying in the form of site grids and rectangular trenches. Setting out grids is dealt with in Chapter 2. A rectangular trench is very similar to a small part of a site grid and the same techniques can be used to set it out. Early site experience will also include levelling, the principles of which are described in Chapter 5. This is a topic which often causes most trouble to the beginner, so it deserves careful study. Usually the day to day levelling on an archæological site does not call for much movement of the instrument. This allows the basic principles to be seen without complications.

The next stage in surveying experience will usually involve surveying existing features, topography, earthworks, remains of buildings and the like. Here a framework of triangulation or traverse will be required as described in Chapter 3. When first faced with the responsibility of surveying something you may feel rather at a loss as how to start. The secret is to choose a framework and put in pegs to mark it out. Then relate the detail to the framework as described in the same chapter. As earthworks are particularly important in archæological and other field-work, the whole of Chapter 6 is given over to them.

Chapter 4 outlines the use of the plane table both to pick up local detail and for more extensive surveys. With practice and skill the plane table can be used very effectively. Too often it is ill-used and the operator hopes that deficiencies in technique will be compensated by sketching in by eye to make it look right. Alternatively it degenerates into little more than a portable drawing board and plays little real part in the surveying. As a field instrument it is rather cumbersome and can be very trying under adverse weather conditions, but its simplicity and robust construction make it very useful in the field, especially for quick surveys of fairly large areas.

Sooner or later the theodolite will be met in surveying work. A generalised account of the instrument is given in Chapter 7. Theodolites all follow the same basic principles, but the makers adopt different means to implement them, and there are differences in the various models available from a given manufacturer. A little experience with one or two models will quickly show you what to look for when using a strange instrument. The theodolite gives a very accurate means of setting out a right-angle, or any other angle. It is also excellent for setting out a very long straight line. In most archæological work this is about all it is used for. This is a pity because it can be an instrument of great power. When used as a tacheometer, as described in Chapter 8, it can be used to record detail with remarkable speed, efficiency and flexibility.

Chapter 9 introduces some calculations which extend the potential of the theodolite. The small effort required to master these calculations will be repaid many times over in the extended range of theodolite work which they make possible, and the simplification of what would otherwise be extremely laborious processes.

The major advances in surveying over the last few years have been due to the use of electronic means for measuring distances and computers for recording and processing data. What was specialist equipment a short time ago is now more widely available, and more commonly met with on site. A general account of its use and its effect on surveying is given in Chapter 10. No attempt has been made to discuss the fundamental theory of the equipment. This would take us outside the scope of this small book.

Surveying may be divided into two main types of activity. The first of these is concerned with finding the position of objects. We might want to find the position of surface features such as walls or column bases, or we may wish to locate a skeleton, a coin or some other find at the bottom of a trench. On a larger scale we might wish to record a whole site such as a major earthwork, camp, or area of countryside including

1: Introduction to Archæological Surveying

roads, hedges and streams. The other side of surveying is concerned with laying out grids and trenches to control work on sites and mark its limits, or else to serve as a framework from which features may be located. In the first type of surveying we have to accept what is there and make the best record of it that we can. In the second type there is a certain element of choice and we decide within limits what we shall do. However, in all surveying the same basic principles are used whichever type of surveying we are engaged upon.

Similarly the two types of surveying are both directed towards putting a plan or diagram on paper as a representation of what was found or what was done. We should always bear this in mind when doing survey work because it will help answer any questions that may arise. First of all it answers the question 'What shall we do?' Try to imagine how you will set the work out on a drawing and what measurements you will need to do this. Secondly it helps us to decide how accurate we need to be with our measurements.

1.2 Accuracy

There is no point in taking time and trouble to measure something to exceptional accuracy if it is impossible to plot the work to a corresponding degree of accuracy. For example with some trouble it might be possible to measure the width of a wall to the nearest millimetre, but on the plot one millimetre scaled down may come out at one hundredth of a millimetre, a quantity that no one could plot! We shall have more to say about this later.

It is important to cultivate neatness and tidiness in your field notes. They should be clear, legible and understandable not only to you but to other people. Bad field notes generate errors. Extensive fair copying is wasteful of time and in itself can be a source of errors. Working under difficult and dirty conditions is not an excuse for scribbled and grubby work, and it may force you back into those same conditions to verify doubtful or meaningless notes.

Always check your work and wherever possible do it by a different method or routine than the one used the first time. It need not be an elaborate check and in some cases can take the form of just looking by eye to see if the things are in certain relative positions. The beginner often seems anxious not to have mistakes in his work discovered, but the experienced surveyor knows how easy it is to make mistakes and likes to discover them before his work leaves his own hands. Unsuspected mistakes carried forward into the future will not only cause embarrassment but can cause an inordinate amount of trouble and waste of time.

1.3 Scale

As was said above the end result of all surveying is to make a plan on paper of what was in evidence on the site. This means that what we draw on paper will have to be many times smaller than the reality and this is what we mean by scale. This ratio of real size to drawn size will depend upon the particular reality we wish to depict, the size of paper we have to work on, and the purpose of the plan.

When Imperial measurements were used scales were often given in the form '1 inch = something', thus for the detailed plan of a trench a scale of 1 inch = 1 foot might have been used. This could also be expressed as a ratio, and these same figures can be stated as a scale of 1:12, or a twelfth. When using metric units scales are given more often in the ratio form, for example 1/100.

Comparison	of	Scales
Imperial	Imperial ratio	Rough Metric equivalent
1in = 1ft	1/12	1/10
1in = 4ft	1/48	1/50
1in = 8ft	1/96	1/100
1in = 10ft	1/120	1/100
1in = 20ft	1/240	1/250
1in = 50ft	1/600	1/500
6in to 1 mile	1/10,560	1/10,000
25in to 1 mile*	1:2,500	1:2,500

* This is the scale of large Ordnance Survey plans. The ratio is exact; the 25in to 1 mile is an approximation.

Table 1.1 Comparison of Scales

It is important to consider the question of scale before starting survey operations, because this is one of the factors which will determine how accurately we must work. Whoever is in charge of the work should clearly state what plans are required and at what scale. It is also a

1: Introduction to Archæological Surveying

good idea to adopt a standard size or sizes of sheet for all drawing work, as it simplifies the storage and sorting problems.

If a scale of 1/100 is used then a normal scale rule, marked in cm and mm, can be used. 1 cm on the scale will be equivalent to 1 m and it is easy to use. Similarly for scales of 1/1000, 1/10,000 etc. and scales of 1/50, 1/500 can also be used with little difficulty. However scales of 1/25, 1/250 begin to be awkward unless you have a properly divided scale, so think of what drawing equipment you have before reaching a decision. Also be wary of making snap decisions about having enlargements or reductions to say one quarter or similar; they may give trouble later.

Adopt a consistent policy right from the start and stick to it.

Chapter 2

Rectangular Grids

2.1 The Site and Grid

To get some idea of what we mean by 'gridding' a site you can think of the operation in miniature. Suppose that we have an aerial photograph of the site and that we have a sheet of plastic drawing film which has graph lines printed on it. Now we wish to fasten the sheet of graph paper firmly to the aerial photograph in such a position that it will be most useful to us in our work (Fig. 2.1). On the real site we cannot rule giant lines all over, but we can mark the intersections of the main lines of the grid, and this we do by driving in stout pegs at the proper points. Now we can imagine the grid lines joining these pegs in two directions at right-angles all over our site. But don't forget the smaller grid-lines, like the sub-divisions on the graph paper, they will run all over the site too, but we do not mark all the intersections.

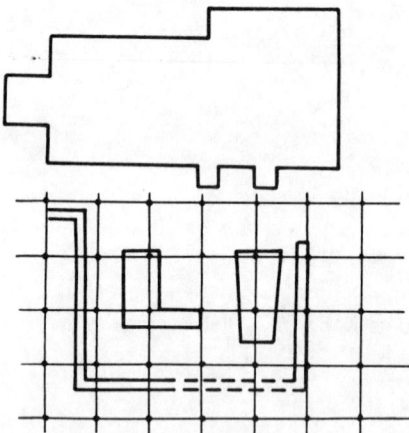

Figure 2.1 A grid superimposed on a plan

Returning for a moment to our sheet of transparent graph paper, we can move it in any direction we please until eventually we pin-point one of the main intersections. Now we can rotate the graph paper about

2: Rectangular Grids

that point until we get the lines running in the right direction and then we can finally fix the two together (Fig. 2.2). This is what we usually do on a site with the grid. First we fix a main grid origin and then decide how the main lines will run through it. From these we can set out the other main intersection points.

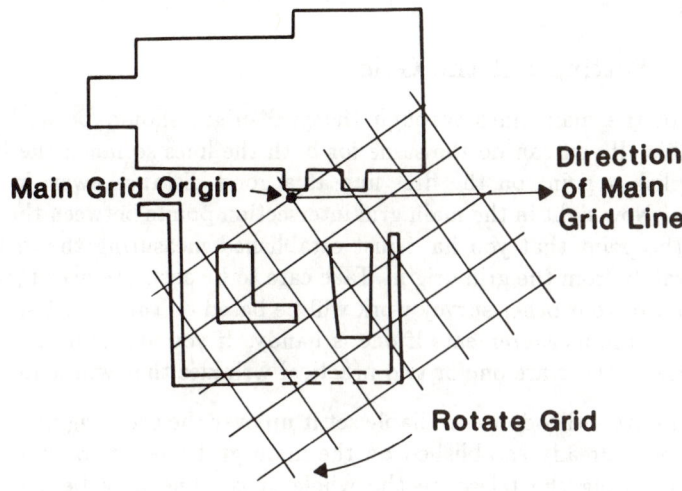

Figure 2.2   Locating grid on plan

2.2 Choice of Origin and Axes

The choice of position of the main origin and the direction of the main lines are very important and careful thought should be given to settling them. Walk around the site, study it and try to imagine your future surveying problems. At first of course you will not have much experience to guide you, but let us consider some of the factors to keep in mind.

First of all consider whether the work you are engaged upon will be continued some time in the future; will other people come to continue your work? If there is the least likelihood of this, then play for safety. What this means is that the main origin should be as nearly permanent as possible or should be capable of being replaced by means of simple measurements taken from well defined points which are themselves on permanent features. If there is a concrete pavement, very large block of masonry or other solid surface in just about the right position to suit the main lines then put your origin on it. Mark it by scratching the surface, cutting a fine X with a chisel or some such means. If you can get them, road nails are excellent markers. Their proper purpose is to fasten to the road surface those tubes which are used to count traffic. They go into mortar joints particularly well. Be careful when driving them in as a badly directed blow may make them fly sideways. If no suitable hard surface exists then knock in a stout peg and drive a nail into the top to mark the exact spot. Leave the nail sticking up about 5mm.

2.3 Setting out the Grid

One of the main lines through the grid origin should be as long as possible. If you can do the same for both the lines so much the better. Establish a point on the first line at a good distance away from the origin. Now sight in the main grid intersection points between the origin and the point that you have just established measuring the distances accurately from the grid origin. Take care to be accurate over this work as a lot of your other survey work will be based on the grid. Use a steel tape for the measurements if one is handy. If you are going to do this effectively there are one or two practical wrinkles that will help.

If you have a theodolite available set it up over the grid origin and sight the point already established on the main grid line. Now by depressing or raising the telescope the whole of the line may be accurately viewed and pegs put in where necessary. However if you do not have a theodolite it is still possible to do the job using the naked eye. The same general principles hold good and used sensibly the eye is a good surveying instrument.

For sighting over medium to long distances ranging poles are used. These are made of wood, metal or plastic, about 2-3cm in diameter, 2m long and coloured alternately in some combination of black, white and red. Over short distances arrows may be used for sighting. Arrows are made from heavy gauge wire and are about 30cm long. Stout metal tent pegs make a very good substitute. The only disadvantage to the use of arrows for sighting is that you have to lie down to sight them properly, but they can be sighted in with remarkable precision.

Before sighting a ranging pole or arrow inspect it to see that it is truly vertical and correct it if necessary. In the process of sighting try to use the lowest parts of the ranging poles; these will be nearer to the points on the ground which mark the correct positions and so will not be affected so much by any slight inclination of the pole.

Fix a ranging pole at the far end of the line to be ranged in. Then fix another one at the near end of the line. Stand close behind the latter and look towards the far pole. Some people get worried about

2: Rectangular Grids

this because the nearer pole looks blurred. But if you move your head slightly this pole will cut off your view, so your eye can only be about the width of the pupil to one side (of course the other eye is shut!). As you direct your assistant (Fig. 2.3) to move the intermediate pole to be fixed on to the line, you will begin to wonder whether it is at last in its proper position. At this stage move your head to sight along the other side of the nearer pole. The view from either side should be the same, showing the intermediate point to be fixed and the more distant point in the same relative positions (Fig. 2.4). This sounds more complex than it is. With experience and practiced assistance it can be done quite quickly. At first you will be slow but speed will come later.

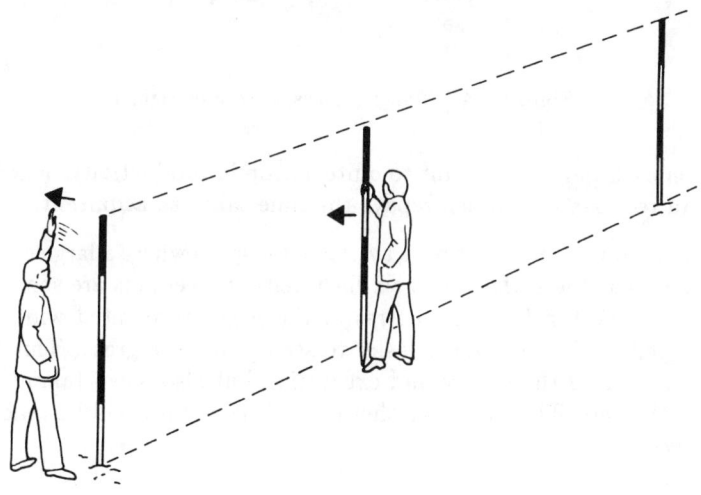

Figure 2.3 Ranging a pole onto a straight line

In fixing the main grid intersection points along the line through the grid origin the procedure will be to drive a peg in at the right position and then knock a nail into the peg. Locate the peg position using a ranging pole and then fix the nail using an arrow to sight to if the distance is small, otherwise use a ranging pole held on top of the peg. In fixing the distances always keep the steel tape as nearly horizontal as possible and again keep it low down so that the measurement is between the real points, not near the top of the ranging poles.

Setting out grids and similar work means driving in a lot of pegs. Sixty or more pegs in a day is not exceptional. Unless it is done efficiently and quickly a lot of time can be wasted; just one minute longer than neces-

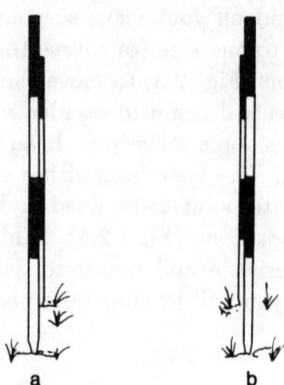

Figure 2.4   Ranging poles, as seen by sighter

sary on each peg can amount to quite a drop in productivity. Knocking pegs in requires skill which repays the time taken to acquire it.

It pays to have some short pegs available for spots where a large stone or rock lies near the surface, and six-inch nails or steel pins are sometimes useful. If the top few centimetres of the pegs are painted white or a bright yellow they are much easier to see against the grass. This helps not only during the survey and excavation but also when taking them out at the end. Then with another coat of paint they will be ready for the next site.

2.4   Setting out a Right-Angle

We are now at the stage where we have one main grid line set out which passes through the grid origin. Incidentally the grid origin does not have to be at the end of the main grid line, it may be anywhere along its length. Now we are faced with the task of setting out the second main grid line at right-angles to the first. If you are working with a theodolite this will be set up at the origin already, and so all you will need to do is to turn off 90° on the theodolite scale. Read Chapter 7 on Theodolites if you wish to know more about setting up and reading angles. However, if you do not have a theodolite you can set off the right-angle accurately using tapes, pegs etc.

Probably the easiest way to set out a right-angle at this stage is as follows. Choose one of the pegs which is set part way along the main grid line (Fig. 2.5). If the main origin is part way along the line then use that position. Now select two pegs on the main grid line which are

2: Rectangular Grids

on opposite sides of the chosen peg and the same distance from it. Take two tapes and fix the end loops over the nails in these two pegs. Now pull the two tapes taut and adjust your position so that the two tapes intersect at the same reading but at a point well away from the main grid line. The intersection point lies on the line which is at right-angles to your main line and which passes through the point you chose. Fix a ranging pole at this point or mark it more permanently with a peg and nail if you wish. When using this method try to use the greatest length possible on the two tapes and arrange things so that the tapes lie at about 45° to the main grid line.

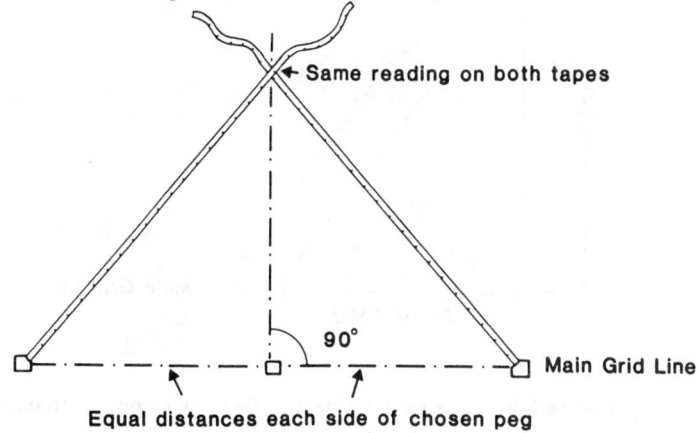

Figure 2.5   Setting out a right-angle using tapes

A very ancient but quite respectable way to set out a right angle is to make use of the fact that if the lengths of the sides of a triangle are in the ratio 3 : 4 : 5 then the angle opposite to the side of length 5 is 90°. Notice that the lengths of the sides need only be in that ratio so that if we multiply all by some number, say 6, we can use a triangle whose sides are 3 × 6, 4 × 6, and 5 × 6, that is 18, 24 and 30 metres. At a distance of 18 from the main origin and lying on the main grid line fix a point, either using an arrow or a peg and nail (Fig. 2.6). Measure from this point a distance of 30, and at the same time measure a distance of 24 from the main origin. Pull the two tapes taut and put in a marker at their intersection. This marker lies on the line at right-angles to the first main grid line and which passes through the origin. We have outlined this method using two tapes and this is to be recommended. It is possible to take the two measurements using one tape, but this should only be attempted with a plastic tape. Never try it with a steel tape otherwise you may put a permanent kink in it or even snap it. In any case you will not get such an accurate result. Incidentally when using tapes looped over nails remember that it is the outside edge of the loop which is the zero of the tape, so allow a little on to your measurements for the thickness of the loop and half the thickness of the nail. The ratios given above are not the only ones which will fix a right-angled triangle: two other sets are 8:15:17 and 20:21:29.

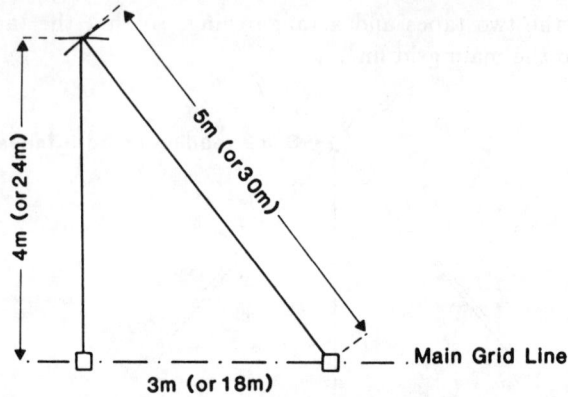

Figure 2.6   Setting out a right-angle using an integer triangle

The method just described has the fault that a point has to be set out on the first grid line at a distance which usually does not fit in with the pegs already set, and the resulting position established on the line at right-angles is likewise at an odd distance. Naturally it would save work if we could use a triangle which had two sides equal to some multiple of the spacing of the main grid intersections (Fig. 2.7). It is possible to do this but the length of the third side will not be an even number of units. This is no great detriment, as when using a metric tape the distance can be set out to the decimal fraction of a unit quite easily. The greater difficulty is in finding the required length which is given by the formula,

$$\text{length of diagonal,} \quad D = \sqrt{2 \times S^2}$$

$$\text{or,} \quad D = 1.414 \times S$$

where D is the length we want and S is the length of each equal side.

2: Rectangular Grids

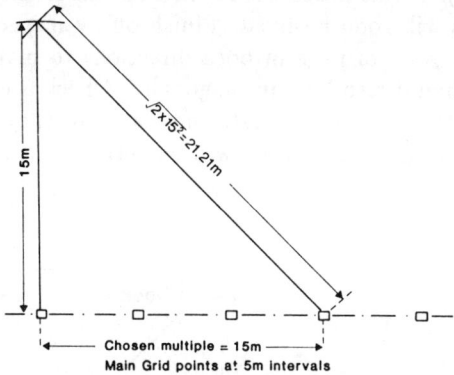

Figure 2.7 Right angle triangle with two equal sides

S	D	S	D	S	D
1	1.414	6	8.485	15	21.213
2	2.822	7	9.899	20	20.284
3	4.243	8	11.314	25	35.355
4	5.657	9	12.720	30	42.426
5	7.071	10	14.142	50	70.711

Table 2.1 Diagonal values, D, for triangles with two sides of length S

Table 2.1 gives values of D for likely values of S. Remember that these quantities are proportional; you can multiply two corresponding entries by any number and the method will still remain true.

Using one of the methods given above, a point on the second main grid line is now known. Using this as a sighting mark, other main grid line intersection points can be set out along this line. We will then be in the position of having two lines at right-angles marked off at regular intervals with pegs or other semi-permanent marks. Our next task is to fill up the remainder of the site with pegs marking the main grid intersections. As long as we have set the first right-angle carefully we do not need to repeat the process since we can put in all the other pegs by reference to our first two lines. Using two tapes, each fixed at a grid point on one of the main lines and pulled taut to the appropriate lengths they will cross at right-angles over the position of a new intersection point (Fig. 2.8). If this process is repeated in an orderly fashion all the required points will soon be fixed. Finish off by inspecting your work. Look along the rows of pegs in both directions to make sure they are in line; some should also line up diagonally. Check a few diagonals for length: even if you cannot calculate the diagonals they should all be at the same distance give or take a few millimetres, so compare the values that you get.

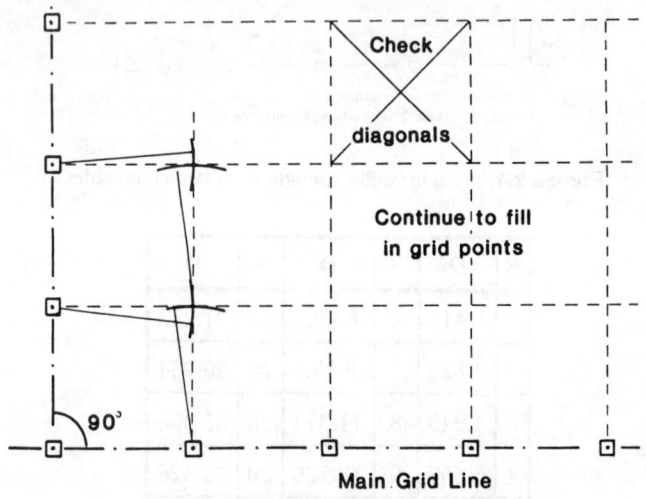

Figure 2.8   Extending the grid

2.5   Uneven Ground

All the above assumes that your site is a flat or gently sloping one and how happy the surveyor would be if all his sites were so. In practice you are likely to come up against bumps and hollows which will call for a certain amount of ingenuity. It may be necessary to throw your grid around the obstacle by setting out more lines at right-angles to the main lines, so that they by-pass the problem area. This will enable you to link up your grid beyond the obstacle and then you can move in on it from all sides. When measuring on steep slopes keep your tape horizontal, even if this means introducing intermediate points (Fig. 2.9).

2: Rectangular Grids

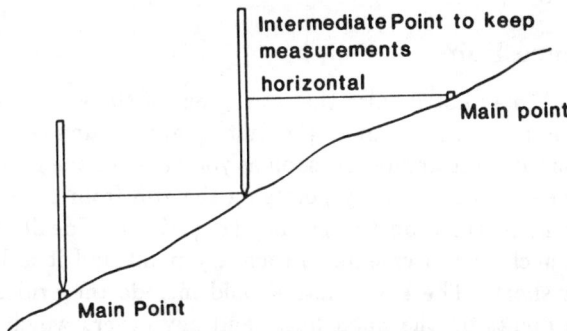

Figure 2.9 Measuring distances on a slope

2.6 Chains

These are sometimes used to measure distances. They are heavier and more substantial than tapes. When measuring through an area covered with shrubs, gorse or other coarse vegetation they are indispensible.

Chains are usually divided into 100 links. The end of every tenth link is marked with a brass tally, the shape of which indicates its value. Going along the chain the tallies at the end of the tenth, twentieth, thirtieth, etc. links are shaped with one point, two points, three points, four points, a round blob, four points, three points, two points and one point. Note the symmetry of the markings about the centre, which for example makes the 40 link and 60 link mark the same. Be careful not to confuse these two. The danger grows less as you move away from the centre; obviously 10 and 90 should not cause trouble.

Some chains are made out of small gauge wire and links are liable to get bent. They are lighter to carry, but inspect them carefully.

Another problem with chains concerns the formation of the loops. In the best quality chains the wire is bent round to form the loops and the two ends are brazed together. Such a loop is virtually unstretchable. Cheaper chains do not have this feature and the ends of the loops are left free. Under rough usage, and chains are intended for use when the going gets tough, the loops can open out a little. At the end of each link there are five loops, so in the length of the chain there are 500 possible points of extension. If each one opens up by 0.1mm the total error will be 5cm. I have known chains with up to about 15cm stretch.

If your chain is not brazed check it carefully against a good steel tape.

Remember that if a line is measured with a chain that is too long the answer that you get will be too *small*.

2.7 Reference Points

One season's work may not require the setting out of the whole grid so this may save you some work, but in the first operation always get the two main lines out at right-angles. As soon as you have got the grid done take reference measurements to key points on the grid from permanent or semi-permanent features on the ground (Fig. 2.10). Ideally there should be three such measurements to each key point and it is better if they are kept short. The key points should include the grid origin and one or two points on the main lines, and any others which seem important to you. These measurements will help you to find your most important points quickly in future years or to replace them accurately if anyone has tampered with them. Make two copies of the sketches showing the measurements, one for you and one for the person in charge to keep with all the papers about the work.

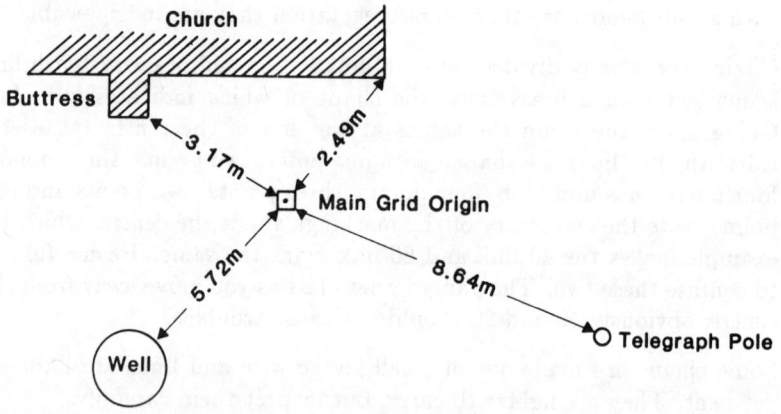

Figure 2.10   Location of a reference point

2.8 False Origin

As described earlier when talking about grids, we usually choose some suitable point as the main intersection from which to lay out the grid. We then need to assign some letters or numbers to the grid lines, so that we can identify particular lines. As a first attempt we might think

2: Rectangular Grids

of marking them +1, +2, +3 etc. in one direction and -1, -2, -3 etc. in the opposite direction. However this can lead to complications and mistakes. Earlier we likened the grids to a piece of transparent graph paper. Now we can think of it as being part of a much larger piece of graph paper and we can start to number the grid lines from whichever spot we wish. Let us choose such a spot down in the bottom left-hand corner of the big sheet, well away from the small portion that we are actually using (Fig. 2.11). Now we can number the grid lines from 0 upwards moving from left to right, and similarly we can start at 0 and number the lines up the sheet. In reality we never go to the point 0, nor do we need to go to it. All that we need to do is to allocate two numbers to the main intersection point from which we started, and make sure that these two numbers are so big that we shall never have to go below 0 on any of our grid lines on any part of the site.

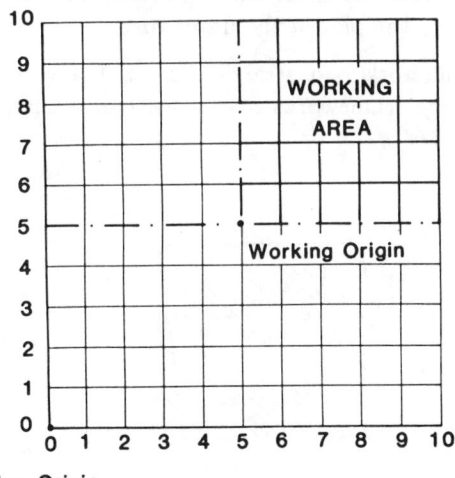

Figure 2.11   Using a false origin for grid references

When quoting a grid reference it is best to give the two numbers, say 4/7, such that the 4 refers to the left-right or west-east direction, and the 7 is the bottom-top or south-north direction. This system has the advantage that it follows the pattern used by the National Grid on Ordnance Survey maps.

Chapter 3

Features and Topography

3.1 Frameworks

If you have to survey the features at the bottom of a trench, a single building, a complex of ruins, or a large tract of countryside, the same general principle applies. First of all establish a framework of survey points and lines, which can be checked to prove that it is correct. Then the details of the features can be surveyed from the framework lines or points and added to the plot of the framework.

The simplest framework is a straight line, but a very common unit is a triangle. Large frameworks may consist of a complex system of interlocking triangles.

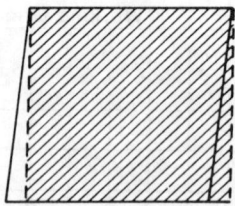

Figure 3.1   Possible errors from assumption of right-angles

In some cases there may be a strong temptation to avoid using a framework, but this will almost invariably be a mistake. For example you may be faced with the job of surveying a building which has four sides, and so you just measure the lengths of all the sides and find that opposite sides are just about equal. If you conclude that the building is rectangular and plot it as such you may be wrong (Fig. 3.1). The better approach is to put a framework of pegs around the building and take such measurements between them as will enable you to plot their relative positions accurately (Fig. 3.2). Then when you know that your framework is true you can pick up the building detail from the framework and plot it with much greater confidence.

3: Features and Topography

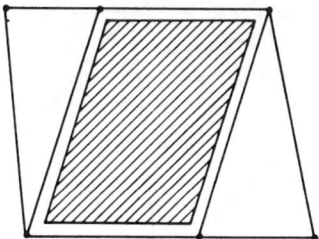

Figure 3.2   Framework around building

The rectangular grid set out over the whole site may form the framework from which you get the detail, but in many cases you will need a completely different frame or a subsidiary frame to deal with particular problems. For this reason it will be helpful to look at possible frameworks. Such frameworks will usually be established by:

(i) Taking linear measurements between the main points.

(ii) Taking angular measurements between the main points.

(iii) A combination of (i) and (ii).

3.2 Linear Triangulation

A framework which depends upon measurement of the distances between the points must form a series of triangles. Such a set of triangles may have common sides or be interconnected in other ways. The points should be so placed that the sides run close to the features which are to be recorded. In deciding which measurements to take between the points do not be content with just the right number to form the figure, take other measurements which are 'surplus'; they will form a useful check on the work (Fig. 3.3). Once the framework has been measured it can be plotted on the sheet by laying down one side in some convenient position and then swinging the other measurements from the two ends to form the triangles which have this line as one of their sides (Fig. 3.4). To do this it will be necessary to use compasses with extension bars or a beam compass to cater for any long lengths. A little experiment with the method will bring home the point that the triangles must have good intersections if reasonable accuracy is to be maintained (Fig. 3.5).

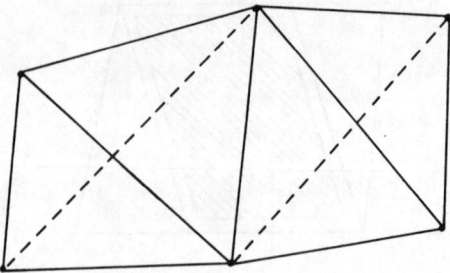

Figure 3.3 Use of check measurements

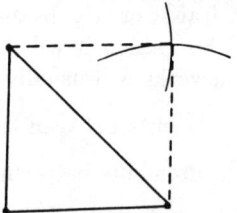

Figure 3.4 Intersection of arcs to locate new point

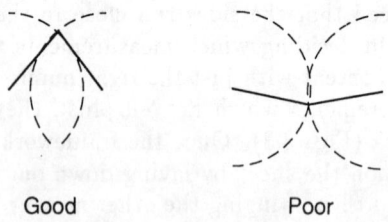

Good   Poor

Figure 3.5 Good and poor intersections

3.3 Angular Triangulation

If all the angles of a framework are measured, then as before the points must form a series of triangles. Since the angles of a triangle should add up to 180° exactly this gives a check on the accuracy of the work.

3: Features and Topography

Rarely, if ever, will the triangles be exact, but an acceptable error should be decided upon. One side must be measured and then it is possible to plot the points using a good, large protractor, starting from this side. As each triangle is formed it will provide a starting point for new triangles. Alternatively the method described in Chapter 9 Surveying Calculations may be used. If surplus angles are measured they help to check the work.

3.4 Traverses

A traverse consists of a set of lines in series, joined end to end. The length of each line must be measured, along with the angle at the end of each line between it and the next line in the series. If the traverse goes round in a circuit back to its starting point it is called a 'Closed Traverse', and the closure forms a check on the accuracy of the work. An 'Open Traverse' does not return to its starting point and no check is possible. Plotting of traverses may be done by protractor and scale, or the points may be calculated as shown in Chapter 9 Surveying Calculations.

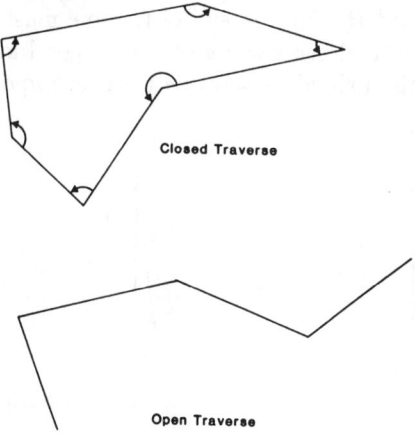

Figure 3.6   Open and closed traverses

3.5 Relating Detail to Framework

The framework of points and lines is the skeleton which must be clothed by adding the detail of the features which we wish to portray. We must therefore consider how we can survey and relate the detail to the framework. The basic problem is to fix the position of a point relative to a survey line. Once we can do this we can fix other important points and draw lines between them to depict the required features.

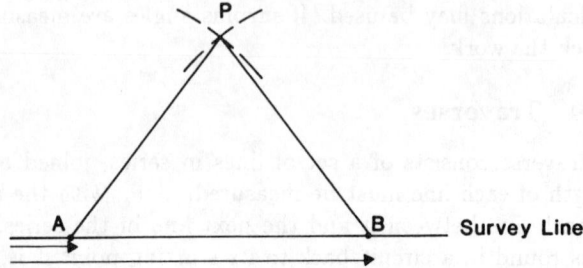

Figure 3.7 Relating detail (point P) to framework (line AB)

A point P can be related to a line by measuring the distances to P from two points A and B which lie on the line (Fig. 3.7). Of course, the distances of A and B from the end of the line must be recorded. With a good shaped triangle this will give a very good location of the point P, and this method should always be used for important points.

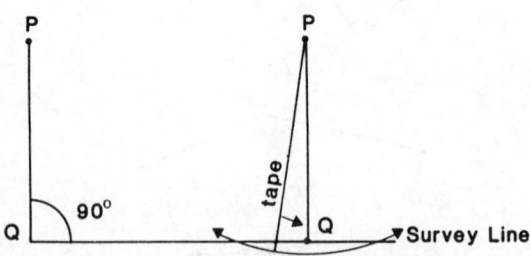

Figure 3.8 Use of 'minimum distance' for obtaining perpendicularity

Another method of locating P is to take a right-angled offset from the point Q on the line. The distance of Q along the main line is noted and the length of the offset PQ is recorded (Fig. 3.8). The problem is of course to decide what is the position of Q which makes the line PQ at right-angles to the main line. If accuracy is not particularly important then the location of Q may be judged 'by eye'. An alternative is to

3: Features and Topography

swing the tape about P as centre, the minimum reading on the tape is the offset distance and the point along the main line at which this occurs is the proper location of Q.

P may also be located by measuring the distance AP and reading the angle PAB. If the distance AP is large, say longer than a tape length, then this gets tedious. To deal with a lot of points like this it may be better to use tacheometry, which is described in Chapter 8.

This is the theory of locating points, but how does it work out in practice? Suppose we look at an example as shown in Fig. 3.9. The line DE forms part of the framework and from this line we wish to locate a wall which runs more or less parallel to it, and which has some kinks and openings in it. Along DE lay out a tape pulled reasonably taut.

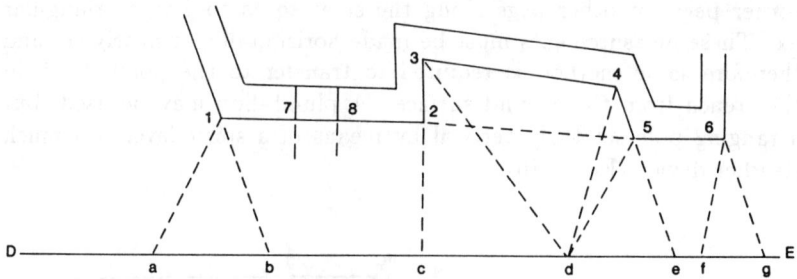

Figure 3.9 Locating points on a feature from a tape line

If DE is longer than a tape length then sight in intermediate points using ranging poles or arrows so that the tape may be put out in the true direction. An alternative approach is to mark the intermediate points $a, b, \ldots f, g$ along DE, which you know you will require, using arrows and then get the distances to them afterwards.

An important principle arises here. When finding the distances of a, b, c, etc. do not measure Da, then ab, then bc etc. Keep the end of the tape fixed at D and measure Da, Db, Dc etc. Each measurement will contain slight inaccuracies and the former method adds them on to each other, while the latter keeps them separate. Try the experiment sometime of measuring along a rough wall with a number of window openings, and door openings. Measure each bit separately and add them up and then check this with one measurement taken overall.

Returning to our example, notice that the little triangles at each end of the wall will fix the points 1 and 6 quite strongly. Measurements 1-7,

1-8, 1-2 and 1-5, will establish important points along the face of the wall and 1-6 should provide a check on the relative positions of 1 and 6. The line 3-2 is continued to c and it is not necessarily at right-angles to DE. A good alternative would have been to make $2 - c$ at right-angles to DE. Whichever way it is done it provides a check on 2; then from the dimensions 2-3 and $d - 3$ the point 3 can be established. Similarly the points 5 and 4 can be fixed. The measurement 3-4 will now provide a check on the location of those points.

So far we have talked about locating things on the surface, or more or less at the same level as the surface. Now we should turn our attention to features which are revealed in trenches, and are some way below the surface. In such cases the grid pegs with nails, from which the trenches have been set out, can be used as the surveying framework. To locate a point in a trench two measurements may be made, either from trench corner pegs, or other pegs along the sides so as to give a triangular fix. These measurements must be made horizontally, or nearly so, and therefore some method is required to transfer to the point down in the trench from the ground surface. A plumb-bob may be used, but a ranging pole set truly vertical by means of a spirit level is a much steadier device (Fig. 3.10).

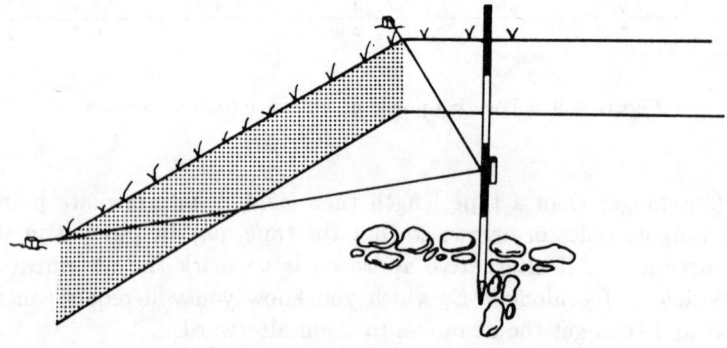

Figure 3.10 Locating features in a trench

This method of transferring from the bottom of the trench upwards to the surface is satisfactory if only one or two points are needed, but if a lot of detail has to be recorded then it becomes tedious and it is much simpler to transfer the surveying framework down to the bottom of the trench. While the top of the trench is most probably set out true and cut accurately, it is dangerous to assume that the bottom of the trench

3: Features and Topography

is equally true. Most of the sides will be cut with a slight batter, so that the sides of the bottom of the trench lie in rather indeterminate positions. This problem may be overcome by adopting the following technique. Set out distances of say 20cm from each corner along two adjacent sides of the trench (Fig. 3.11).

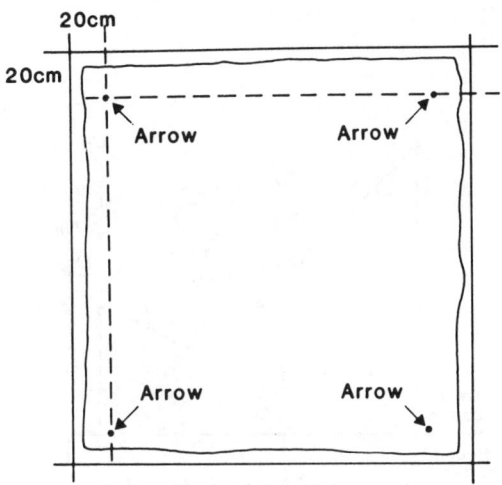

Figure 3.11 Transferring the framework to the bottom of a trench

Put ranging poles at each of these points. Now by sighting across and down, arrows can be placed in the bottom of the trench at the intersections of the lines defined by the ranging poles. These will give a rectangle, the sides of which are each set 20cm in from the sides of the trench at the top. Thus they will normally be well clear of any batter on the sides of the wall and they can form the basis of a little survey framework at the bottom of the trench. Detail work can then proceed down in the trench free from wind or other disturbing influences.

If the trench covers a large area it may be necessary to fix grid points in the bottom. This may be done by using either of the two methods outlined above, or a theodolite may be set over a grid peg at the side of the trench and sighted along the grid. Then measurements can be taken along the line to establish the required points. This will mean transferring the measurements down vertically as before, which may be troublesome. It is possible to do the job without taking any linear measurements by using one theodolite and a little ingenuity, or better still with two theodolites if you can lay your hands on them. The principle of the method is shown in Fig. 3.12, from which it can be seen that intersecting lines of sight from grid points at the sides of the trench are used to fix the grid points in the trench. The angles are used to define the sight lines. The calculations to find the angles are quite simple and will be found in Chapter 9. In sections 9.5 and 9.6 details are also given of similar applications which should make the procedure clear.

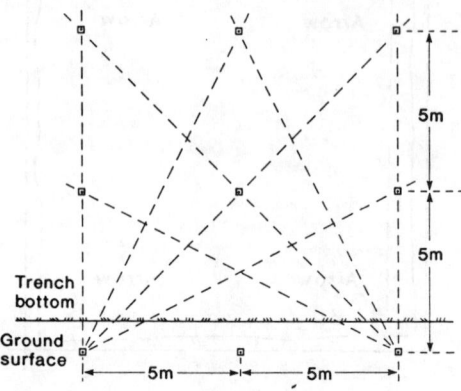

Figure 3.12 Grid in trench

3.6 Optical Square

This is a useful little device for setting out a right-angle. It is held in the hand or put on top of a short pole at about eye level. A specially cut prism gives a view straight ahead and simultaneously a view of objects to one side or the other, slightly above or below the forward view. If two objects line up in the two views they are at right-angles from the optical square.

It is useful for ensuring that long offsets, as described in surveying features, are at-right angles to the main survey line. It may also be used in other cases for checking or setting out right-angles where great accuracy is not required. Unless the vertical axis of the prism is held truly plumb then a greater or lesser error will be introduced.

3: Features and Topography

3.7 Cross-Staff

Similar in purpose to an optical square, a cross-staff employs sighting slits or vanes which are held rigidly to give right-angle sighting lines. It is usually socketed to sit on top of a short pole with a tapered top. Unless the pole is set reasonably vertically the lines of sight will not be true.

The cheapest variety has a simple cruciform shape with the ends of each arm turned up and slit to provide a sight. Unfortunately they are easily damaged, the upturned ends being unprotected and easily distorted by rough usage. This leads to the sight slit becoming a squint in more ways than one!

A slightly more robust, and certainly more elegant version has a hollow, octagonal shaped body about 90cm high and 75cm across the flats. There is a slit and sighting vane in each face, so because of the octagonal shape it is possible to sight at right-angles and 45° to a given line.

It is very useful as long as due regard is paid to its limitations. Like the optical square it must be as nearly vertical as possible when used. The fact that the sighting base is only about 75cm and it is used with the naked eye means that the angular accuracy is not very good.

An excellent account of its use is given in *Fieldwork in Medieval Archæology* by C Taylor (see section on Further Reading).

3.8 Prismatic Compass

This little instrument enables you to take the magnetic bearing of an object or a line. There are a number of types on the market, but the military pattern is the best. They are well made, ruggedly built and will give many, many years of service. All patterns have some form of forward sighting mark, often incorporated in a hinged lid. The rear sight usually swings up into position and the forward mark can be lined up on to the target. At the same time, by means of a prism (hence the name), the reading of the compass card can be seen. When held in the hand bearings can be obtained to an accuracy of about half a degree. If it is supported on a short post a little better accuracy is possible, but the sighting base of the instrument is small so that the accuracy of alignment is never very great.

The prismatic compass can be used very much like a miniature theodolite but remember that its accuracy is much less than that of a theodolite. A compass traverse can be set out, or points located by taking bearings to them from known points. When using the compass the usual precautions should be taken against local interference from magnetic materials, particularly steel and iron objects. Even spectacle frames may be magnetic.

3.9 Connecting to Ordnance Survey

If very great precision is required this can be a long, complex procedure. In such a case work must link to OS triangulation pillars or Reference Points. However work of this nature is beyond the scope of this book. What we shall now look at is how to relate to an OS map with an accuracy compatible with the methods of surveying we have considered.

Study a large scale OS map of the area, either 1/2,500 or 1/10,000. Pick some well-marked permanent features such as corners of buildings, intersections of field boundaries and the like. The minimum number of points required is two, but it is better to have three or more so that there are adequate checks. The points should also be well distributed over the area, not huddled closely together.

Having chosen suitable points survey them in along with the other detail, but be particularly careful and use the most accurate method. For example fix them with linear triangulation if appropriate, and again take extra check measurements to them. If your survey is now drawn to the same scale as the OS map, these points will provide a set of register points for aligning the two surveys. The National Grid lines may then be transferred if required. Alternatively, the OS map may be enlarged to fit the survey scale. Remember that one point is not sufficient, it will merely provide a common point around which the plot may pivot as described in Section 2.1.

3.10 Detail

When surveying and drawing detail use discretion in your work. A good example of this can be found in drawing the detailed plan of a stone wall. It is a complete waste of time to be over-meticulous in measuring up the detail of individual stones. What you should aim to do is to record the character of the masonry work. After all it is fortuitous that you are working on the top of that particular course. At some time there were courses above it and there are still courses below it. If one of these had been open to you the minutiæ of exact measurements to a millimetre would have been far different, but the character of the work would have been the same. This is also an outstanding example of the necessity of measuring from some fixed point at the end. If you measure up each stone individually and try to build up a drawing of the wall bit by bit you will inevitably fail to get the right overall dimension. First of all get the overall boundaries of the wall fixed and then lay a plastic

3: Features and Topography

tape along the length of the wall. Now the individual stones may be measured with a metal or wooden rule and drawn in, but the plastic tape serves as a check on their general position.

Finally do not forget that every plan should have a north point, a scale, the date it was drawn, the site and any other relevant information which will help in its interpretation such as depth, datum, description of level etc. Without such information, in several years' time your plan will be meaningless and all the work entailed in drawing it may be wasted, and possibly the whole of the project rendered void.

3.11 Plotting

Before starting work you should decide where you will do the plotting or drawing. For a beginner it is a wise precaution to do a reasonably finished plot either on site or close by. Then the plot can be compared with the physical reality. It is worth while going to some trouble to arrange that your first surveys are close to your permanent base so that the sites can be easily visited to check the final plot. But if you are going to be a competent surveyor you must gradually phase out this stage. Treat it as a crutch which you are determined to dispense with eventually. I know that many people advocate such checks but there are two very good reasons why you should learn to dispense with them.

First of all good finished plots can be produced more easily and quickly in a proper environment which is not usually available on or adjacent to the site. On site it is quicker to record observations and make detailed field notes than it is to plot. To make best use of the time on site it is better to devote it to making a full and proper set of field notes, and remember time on site is usually more expensive than time at the permanent base. Check the work before you leave the site, just as carefully as if it were a finished plot. Make sure that your notebooks are sufficiently detailed for you to plot everything you see and need without ambiguity and misrepresentation. Walk round and think carefully. At the intermediate stage do this even if you are going to return to check your final plot. It is a good discipline and will stop you falling into shoddy methods of booking, and persuading yourself that you need not record detail because you can remember it - and in any case you think you will put it all right when you come back with the full plot!

The second reason for working towards this approach is that if you continue to do much surveying there will come a time when it will be out of the question to go back. Unless you have trained yourself as I suggest you will either have to turn the job down or be in danger of making a mess of things.

3.12 Drawings

Very often graph paper is used for plotting plans on the grounds of ease of use, since the framework is already set out on the paper. It is not generally appreciated that paper stretches different amounts along the two axes, and if a sensible length, say 40-50 cm of graph paper is taken it will often be found that measurement by a good quality scale rule does not agree with the printed lines. This means that detail traced from a plot on one piece of graph paper may not fit correctly on to another drawing.

The safer procedure is to plot on to plain paper using an accurate scale. Better still is a plot made directly on to good tracing paper, tracing linen, or drafting film. Reproduction is then much easier.

It is possible to buy drawing film on which graph lines have been printed with high precision. Use this if you want convenience, allied to accuracy, but it is expensive.

It is a good idea to use standard sizes of paper and stick to them. Ideally one size sheet should be used, but this may not be convenient for both plans and sections. However, if the sheets are made to fit the size of the work some will be large and others will be small. This leads to great inconvenience in storing, sorting and handling, and can lead to loss of small sheets among masses of larger sheets.

There may be difficulties where work spreads over two or three sheets, but no more than one finds in using Ordnance Survey maps. In any case those two or three sheets can be much more easily stored and looked after, and following repeated use will be in a much better state than a large roll which has had to lie around because none of the usual storage facilities would accommodate it.

Chapter 4

The Plane Table

4.1 Equipment

When the operator has a good knowledge of the principles and practice of surveying, with a lively awareness of the sources of error and their minimisation and good drawing technique, the plane table can produce good results. In Britain the weather can make it a trial to use. Beginners should approach it with care and respect.

Unlike most other pieces of surveying equipment the plane table is not complete in itself and it requires a number of accessories to go with it so that it may fulfil its function. On the other hand when the field operations are complete there should be little further office work to do, as the plot produced in the field should only require a little tidying up for a fair tracing to be made.

The plane table itself consists of a flat drawing surface which may be attached to a tripod. The accessories required are an alidade, a spirit level, a plumbing fork and plumb-bob.

The alidade is a sighting device combined with a straight edge. Some are very simple affairs with sighting vanes which fold down for storage. Unless they are stoutly made the vanes are liable to become distorted under hard service conditions. Other alidades are fitted with telescopes which have stadia wires so that distances can be found by tacheometry. Usually the telescopes are mounted so that their line of sight lies directly above the straight edge which is used for drawing lines. Some of the best quality alidades have a parallel-rule motion attached to the straight edge, and this is very useful.

The spirit level should be robust and of good quality. It does not need to be as sensitive as those on a theodolite as the plane table top will deflect easily and with too sensitive a level it would take an unjustifiable amount of time to level up. On the other hand a cheap insensitive level should not be used. Often some form of level is incorporated in the alidade. If the alidade is a cheap, simple affair do not expect too much of the circular bubble that is probably fitted to it.

The plumbing fork is a strange looking device, which is made out of a solid metal or wooden section and which fits on to the plane table as

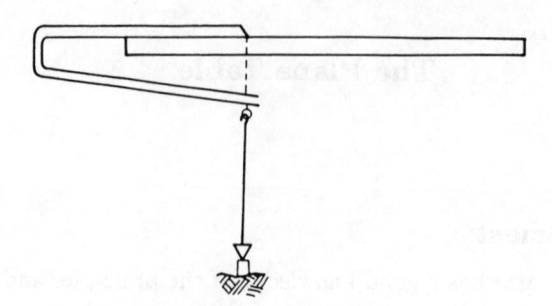

Figure 4.1   The plumbing fork

illustrated in Fig. 4.1.

The plumbing fork slides on to the surface of the table, and the plumb-bob is suspended vertically below the point of the plumbing fork. It enables us to set a point marked on the plot on the table, vertically above the corresponding point on the ground. It should be long enough to allow the point on the fork to command any point on the table. It has been known for people to hold a stone against the underside of the table at the estimated position of the plotted station. Then they release the stone and if it hits the peg they consider the instrument in the right position. This shows that they know what should be done but have not been provided with the right equipment and may not even know that it exists.

For plane tabling in Britain always use a plastic draughting film, well sealed down all around the edges. This makes the drawing surface rain and wind proof. Ordinary paper quickly disintegrates with a little wind and rain.

4.2   Outline of Intersection Method

In this method a base line is set out and measured. The plane table is set up at one end of the base line and the base line is drawn in. From the point on the plot representing the end of the base line, lines are drawn in the directions of the features which it is required to fix. Then the plane table is moved to the other end of the base line and the base line on the table is aligned with the base line on the ground. Again lines are drawn from the second end of the base line in the directions of the required features. Where these lines intersect gives the positions of the features. Now let us look at this in more detail.

4: The Plane Table

4.3 Intersection Method Procedure

First of all put in the base line pegs A and B (Fig. 4.2). Then decide on the scale of the plot. This will be limited by the size of the plane table. You will not be able to make a plot bigger than the table will accommodate. This sounds obvious, but it requires some thought at this stage and you should make a careful estimate of the area that you are going to cover. It is very frustrating to find, after a lot of work has already been done, that some of the plot disappears off the edge of the table.

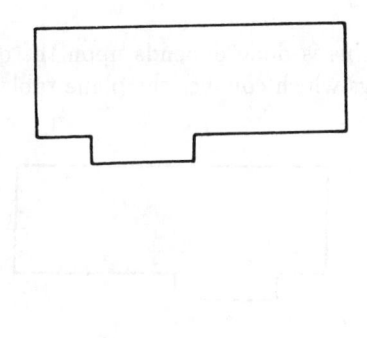

Figure 4.2   Site with baseline

Put a point a on the plot to represent the point A on the ground and then set up the plane table over A, centring the point a on the table over A. At the same time ensure that the table is aligned so that the plot is going to come in the right place. Now level the table. The next stage is to plot the line ab. Keeping the straight edge just touching a, move the alidade to sight the point B and then draw in the line. Measure ab to scale to represent AB and plot b. We can now start to pick up the features that we wish to record. To do this still keep the alidade touching a and sight to the first point. Now draw in a line lightly along the straight edge and give it some identification mark. It is important to make the line a light one and to give it a clear identification. If there are a lot of features we are going to have a mass of lines which will have to match with their corresponding ones from B, and most of which we shall have to rub out in the end. An alidade with a parallel rule attachment makes the operation of drawing the lines easier. The alidade can be kept a little to one side of point a while the object is being sighted. Then the straight edge is moved parallel to the sight line until it touches a. There will be a slight angular error introduced, but as long as the parallel rule motion is kept small this should not be very significant. Once all the required lines have been drawn at A (Fig. 4.3) the plane table can be moved to B. In setting up at B we have to fulfil three conditions, which are as follows:

(i) the table should be level,

(ii) point b should be centred over point B,

(iii) the line ba should be aligned on BA to give the proper orientation of the table.

The way in which this is done depends upon the quality of the plane table and the fittings which connect the plane table to the tripod.

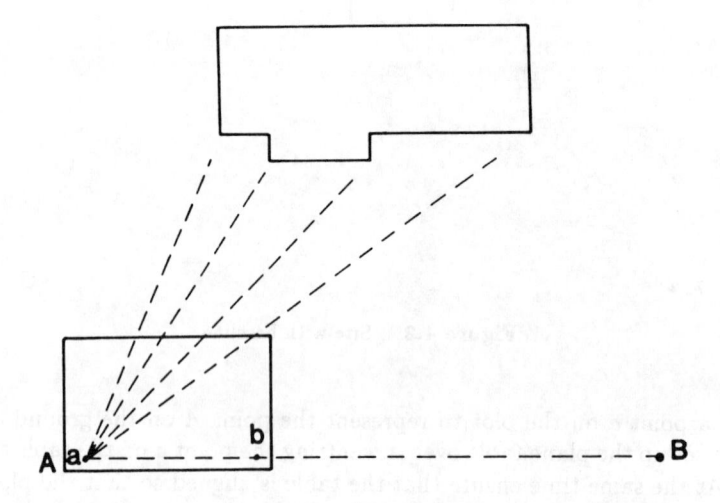

Figure 4.3 Plane table at A

Some plane tables have levelling screws or a ball and socket joint. In addition they may have a clamp and slow motion screw so that the table can be rotated under close control. The main difficulty stems from the fact that any alteration in alignment will affect the centring of b over B and vice versa. There is no hard and fast rule, but a good procedure is to get the table roughly levelled, roughly centred and roughly aligned. Then by moving the legs put the centring nearly right, then put the alignment nearly right. Work at these two until things are reasonably satisfactory, and then level up. This may disturb the alignment. If so

4: The Plane Table

go back and work at it until the three conditions are satisfied. This is not as straight-forward as setting up a theodolite; it requires knack and patience. You will notice that in the description of setting up at A the base line was drawn on after the table was set up. This got rid of one of the conditions, which makes life a little easier, but that dodge can only be done at the first station occupied; thereafter you must obey the alignment condition to keep the table in its correct orientation. Once the table has been properly set up at B observations are taken to the points of detail already sighted from A. Each new line from B will 'intersect' with the corresponding line from A to give the location of the point (Fig. 4.4). When the points are joined up they will define the feature being surveyed.

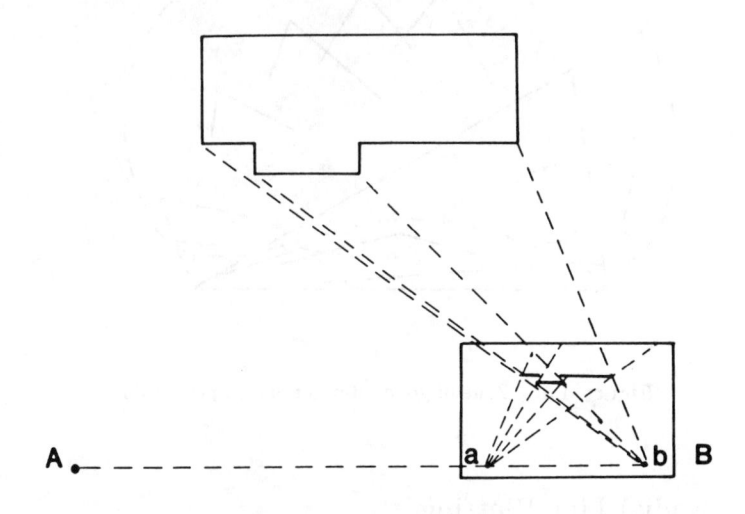

Figure 4.4 Plane table at B

4.4 Development of Method

If the features are quite close then it is possible to measure some of the lengths and check them against the corresponding lengths on the table. Even without such a direct check it is possible to look at the features and compare them with the plot to see that it is generally a faithful reproduction.

The length of the base line controls the area which can be covered accurately by plane tabling. As the angle of intersection of the rays gets further away from 90° in either direction so will the accuracy fall off. Fig. 4.5 shows the area that can be covered to an acceptable degree of accuracy from two points on a base line. Outside the rough boundary marked by the dashed line the angles of intersection will be too small or too large. If a large area is to be covered from a single base line then it must be of sufficient length to provide good intersections. Alternatively new lines may be laid out, connecting with the original one, to form a traverse (Fig. 4.6). Such an arrangement will provide good accuracy for close features and also give widely spaced stations for fixing distant points.

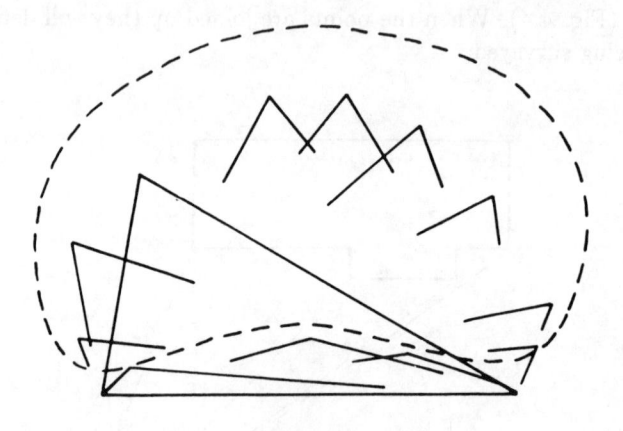

Figure 4.5   Zone of good intersections on plane table

4.5 Radial Line Plotting

In this method the alidade is sighted towards a point of detail and the distance from the table to the point is measured. The length is then laid out to scale along the sight line. Measurements to near points may be made by tape, but as the distances get greater this form of measurement is tedious and slow. If it is intended to adopt this method on anything other than very limited areas it will be best to use a telescopic alidade fitted with stadia wires. By this means distances can be found quickly by tacheometry. A small electronic calculator will facilitate the reduction of the results for plotting.

While radial line plotting has been described separately from intersection there is no reason to keep them separate in the field. On a particular site the two methods can be combined. With a large team one group can be measuring distances to near objects while another person is working

4: The Plane Table

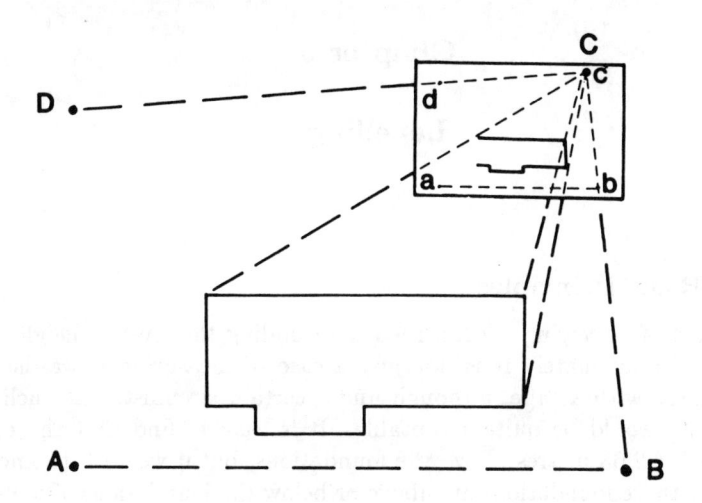

Figure 4.6 Use of a plane table traverse

his way around the more distant points for location by intersection. A skilled person on the table will be able to cope with the two concurrent sets of activities.

Naturally any assistants who are going to be working at some distance from the station must be intelligent, interested and experienced in the requirements of the plotter. If they miss points at some distance away their journeys to and fro to make good deficiencies will waste a considerable amount of time.

Chapter 5

Levelling

5.1 Basic Principles

This part of surveying is concerned with finding the level or height of things. Unfortunately it is not just a case of measuring upwards or downwards with a tape, although under certain circumstances such a procedure would be quite acceptable. It is easy to find that the top of a wall is 2.63 metres above the foundations, but if we wish to know whether those foundations are above or below the foundations of a wall at the other side of the site we shall have to use a surveying instrument known as a level. It is very difficult to judge differences of level by eye, so never attempt it; always use the proper instrument for the job.

If we consider a circular cylindrical bowl made out of glass or perspex which is part filled with water (Fig. 5.1), the water surface will assume a horizontal or level position and we can look along the surface of the water in any direction. This means that we have fixed a horizontal plane from which we can measure downwards (or up if we like) by sighting along the surface to measuring staves that we can locate at different points in the surrounding area. Naturally to do this would be very cumbersome, but the Roman method of levelling was not very far removed from this basic idea. They used some sort of trough, possibly with floating sights at each end, or a tube full of water connecting two vertical glass tubes. They certainly sighted to a staff and took readings from it very much in the modern manner.

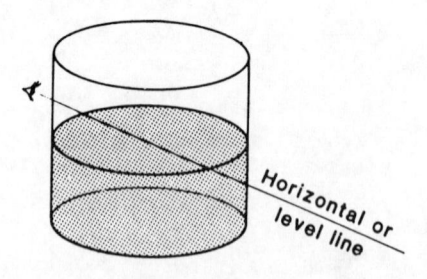

Figure 5.1   Level plane defined by a bowl full of water

5: Levelling

Returning for a short time to our glass bowl, you can appreciate that it will define different horizontal planes depending upon the height at which it is set initially. Thus we can imagine any number of horizontal planes all passing somewhere above the location of the bottom of the bowl and stretching in all directions, like a pile of gigantic and invisible sheets of plywood. The levels that we use in surveying do just the same job as the bowl of water but are rather drier to use. When set up properly the cross-wires in the telescope point along a level line, and if we rotate the instrument about the vertical axis then that level line moves round the particular horizontal plane that the instrument is set to (Fig. 5.2). Incidentally it would be very difficult to set the instrument to a given horizontal plane, so we usually set it up in some convenient place and then by taking a reading on a staff held on a point of known elevation we can find out which horizontal plane we happen to have picked, but more of this later.

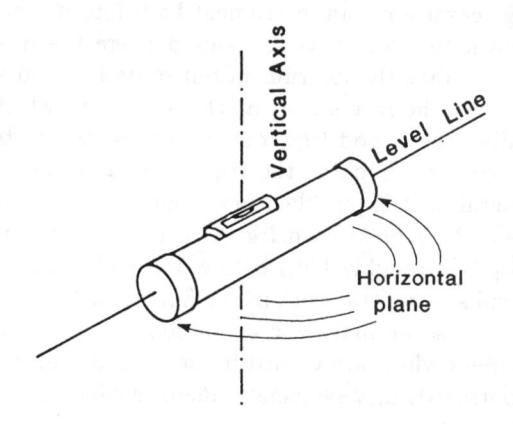

Figure 5.2 Use of a spirit bubble to define level plane

5.2 Types of Levels

It is time that we looked a little closer at the instrument which we have so far described as a level and find out what its characteristics are and what is the best way to use it. There are several types of level in use, but the three which are most likely to be available are the dumpy, quick-set and automatic. A certain mystique has grown up around levelling and seems to be particularly associated with the words 'dumpy-level'. Anyone who knows how to use a 'dumpy' acquires a certain aura. Do not let this put you off, keep in your mind the picture of the horizontal planes and levelling becomes relatively simple. The mechanics of setting up the level is a technique which becomes easy with practice. The quickset level is, as its name implies, rather quicker to set up, but it is not thereby a better instrument to use for all types of levelling. However, on site you will have to use what you have got.

The telescopes of most surveying instruments, with the exception of automatic levels, give an inverted image of the field of view. There are sound technical reasons for this, but you will probably find it confusing at first. Take time and take care and eventually it will cease to trouble you.

5.3 Setting Up - Dumpy

Surveying instruments fit very snugly into their boxes, so study the way the instrument sits in its box before removing it. This will save endless trouble when trying to pack it up again, and at this stage a further warning: never force an instrument back into its container, nor force the lid down to close it as this can damage the instrument or break the bubbles. Take the instrument out of its box and attach it to its tripod. Now set the instrument at the spot from which you wish to work, spreading the tripod legs, but do not as yet try to get them firmly into the ground. Look at the tripod top and judge whether it is nearly horizontal, if it is out then move one or two of the legs until you are satisfied. Now press each leg in turn firmly into the ground. If the tripod top was roughly level before, it should finish up roughly level and that will save a little work in the finer adjustment of the level. For this latter process we make use of the footscrews. A dumpy level has three footscrews which are turned by prominent milled heads. You must use the footscrews in a systematic manner which takes far longer to describe than to do.

Turn the instrument so that the bubble-tube lies parallel to two of the footscrews and then turn those two footscrews equally in opposite directions until the bubble comes to the centre of the tube. A useful tip is to remember that the bubble will always move in the same direction as your left thumb; try it and see. Now turn the instrument through 90° and using the third footscrew only, bring the bubble to the centre of the tube (Fig. 5.3).

If all things were perfect in a perfect world the instrument would now be truly level, but alas we must check and rectify if necessary. Turn the instrument back to the original position and check the position of the bubble, if necessary bring it back to the centre by moving both footscrews in opposite directions. Now return to the 90° position and similarly check and adjust. This procedure should be repeated until

5: Levelling

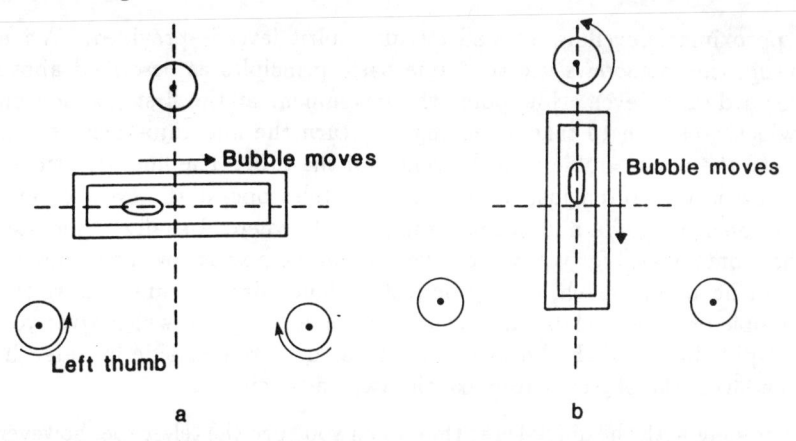

Figure 5.3 Three screw levelling with bubble

you are satisfied with the instrument in these two positions. Next you must turn the instrument through 180° from its initial position. If the bubble comes to the centre of the tube you are home and dry. If not your instrument is out of adjustment, but do not despair, most of them are, so you now proceed with the following routine. Note how far the bubble is away from the central position. Suppose that one end is displaced a distance of 3 divisions marked along the glass towards the eyepiece end. Now mentally halve this distance, which will be 1.5 divisions. Take this position as the proper bubble location and repeat your levelling in two directions at 90°. Finish off with a final check through 180°. When complete the bubble should settle in the same position no matter in which direction the instrument points. This is known as the constant bubble position.

You may think that all those instructions are too much and that by playing with the footscrews you can get the same result much quicker. All professional surveyors use the routine described and when followed through smoothly the instrument should in most cases be out of its box and set up ready for sighting in no more than five minutes. At first it will take you longer, but following the routine is always quicker than any other method.

5.4 Setting Up - Quickset

The quick-set instrument is levelled in two stages. After getting the tripod top set roughly level and legs firmly into the ground the instrument is levelled approximately. This is done by using the three footscrews. Some quick-sets do not have footscrews but are mounted on a ball and socket joint with a locking ring. In order to control this approximate levelling a small circular spirit level is provided. When using the footscrews use the same basic principles as described above for a dumpy level. Now point the instrument at the first position on which you wish to take a reading and turn the fine adjustment screw until the main bubble is in the centre of the tube. The fine adjustment screw is very often just underneath the telescope at the eyepiece end but on some modern instruments may be elsewhere. Usually a quick-set has some device for you to observe the bubble position without moving your head very much, It may be just a plane mirror inclined above the bubble or a more elaborate device such as an eyepiece which gives you a 'split' image of the bubble and brings the two ends side by side. In the latter the object is to make the two ends coincide.

The snag with the quick-set is that when you turn the telescope, however slightly, to view another position you must check and adjust the main bubble, and sometimes you will forget. With the dumpy once it is set up you should be able to point it in any direction, with only the odd glance at the bubble to see that it has not wandered. So you see that if you wish to take a lot of sights from one instrument station the time that you saved in setting up the quick-set may be frittered away in constant bubble adjustments for each sight.

5.5 Automatic Level

The automatic level is very simple to use. Once the instrument is set up with the tripod legs firmly in the ground all you have to do is to bring the small circular bubble within the ring marked on the glass. Then an internal device automatically ensures that the line of sight is horizontal.

5.6 Datum

We now have a knowledge of the equipment used for levelling and so we can look into the ways in which we use them. All levelling is a matter of finding out how far one point is above or below another point, and as a matter of convenience we choose one horizontal plane to be the base from which we shall work. This chosen plane is called the datum plane or simply the datum. For the Ordnance Survey maps the datum chosen was mean sea level, but to establish mean sea level is not quite so simple as you might think. The present mean sea level is based upon a series of observations carried out at Newlyn in Cornwall and all modern O.S. maps work to the Newlyn datum. However on older OS maps the datum was based upon the Old Dock Sill at Liverpool and there is a difference between the two, so always check when doing any serious levelling work from OS maps as to which datum is relevant.

5: Levelling 43

5.7 Bench-Marks

The Ordnance Survey have established a series of bench-marks all over the country and their positions and values are usually to be found on the large scale maps. The bench-marks are usually cut into solid masonry or brickwork (Fig. 5.4). If there is an OS bench-mark near the site then it is very well worthwhile to link your work in to OS datum. This ensures that subsequent workers can do likewise and get a comparison with your work. As you will not want to start from the OS bench-mark every time you need to take a level, it will pay you to establish one or two temporary bench-marks around your site. These can be such points as a permanent piece of masonry, or a stout peg driven into the ground or a brick on end set in a mass of concrete. The points chosen should be definite, recognisable and easily identified, and should be as permanent in character as possible. Level from the OS bench-mark in to your temporary bench-marks and back again as a check. If it is such a long way to the nearest OS bench-mark that it is impossible to do the necessary levelling out and back again, then establish the most permanent of your site bench-marks as your own datum. All this means is that you arbitrarily assign a level to it, to suit your convenience. Calling it 100.00 is a good idea. The principle to adopt is that none of your levels will come out less than 0.00, otherwise you will have minus levels, which become a little messy to deal with and can lead to stupid errors.

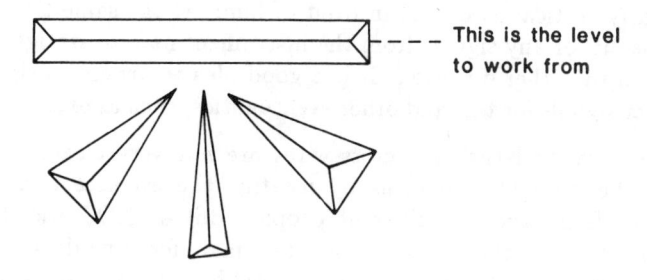

Figure 5.4 Ordnance Survey bench-mark

5.8 Basic Levelling

The operation of levelling always starts from a bench-mark, either OS or a temporary one. The instrument is set up at some point which will give a good sight to the bench-mark, and it is then levelled. The staff is held on the bench-mark so that the bottom of the staff coincides with the point whose level is known. Now the staff is read by looking at it through the telescope and noting the reading against the horizontal cross-wire in the telescope. This value is added on to the level of the bench-mark and the answer is the level of the cross-wire in the telescope. Now the staff can be held on some point whose level we wish to find. Again a reading is taken on to the staff. By subtracting this value from the level of the cross-wire as found above we shall be left with the level of the new point. This simple addition and subtraction of staff readings taken on points of known and unknown levels is the basis of levelling work. Once we have set up the instrument and determined its level by taking a staff reading on to a known point, we can find the level, or fix a level, anywhere within the limits of the length of the staff below the level of the instrument.

The staff should be held vertically. Some authors recommend that the staff should be rocked slowly backwards and forward in the direction of the level. Then the surveyor notes the lowest reading, which will occur as the staff passes through the vertical. I have always found this to waste a great deal of time, especially with beginners. They have problems enough with a stationary staff, never mind one that is giving a varying reading. The person at the instrument can tell if the staff is vertical in one direction because the vertical cross wire in the telescope is an excellent guide. The staffman should stand so that he can judge that the staff is vertical in the other direction. He places himself so that his shoulder points towards the instrument and he keeps the staff as nearly vertical as he can in front of him. At the same time he keeps a lookout for any signals from the instrument man to straighten the staff up in the other direction. It is a good idea to arrange a simple code of clear signals for this and other eventualities, such as extending the staff.

You may find that the cross-wires are not very clear to you. They can be brought into focus by rotating the eyepiece and this setting very often varies for different people. But do NOT use this to focus onto the staff; there will be a separate main focusing device somewhere. Most instruments have short horizontal lines above and below the main horizontal cross-wire. Avoid using these when levelling. The line of sight defined by the intersection of the main cross-wires is called the line of collimation.

Do not attempt to take long sights. It becomes difficult to read the staff accurately at distances greater than 65m. At very long distances the curvature of the earth and refraction affect the results.

5: Levelling

5.9 Booking and Reduction

All the readings need to be recorded and as in all survey work orderliness and neatness are essential if the records are going to be legible and intelligible after the field work has been done. There are two standard methods of booking levels, these being called 'Rise and Fall' and 'Collimation'. Only the latter will be described now, but if you can, find out something about the other. The object of having standard methods is that all people making use of surveying will understand without further explanation. If you write in your own hieroglyphics, in your own conventions, on the back of an envelope, subsequently spattered with mud and coffee, you will have expressed your personality but your surveying work will be highly suspect, and even you may not be able to understand it later on.

An example of a Field Book with a few entries is shown in Table 5.1, the entries corresponding to the illustration in Fig. 5.5.

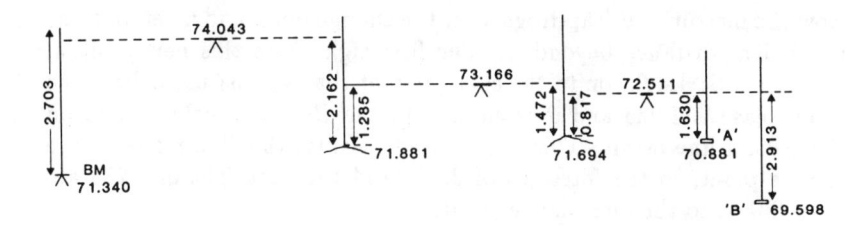

Figure 5.5 Transferring the level

Back Sight	Intermediate Sight	Fore Sight	Collimation Level	Reduced Level	Remarks
2.703				71.340	Bench Mark
1.285		2.162			Change Point
0.817		1.472			Change Point
	1.630				Stone slab A
	2.913				Grave Cover B

Table 5.1 Booking of Levels

Let us look at this booking in more detail. Each time that the level is set up in a new position, a sight has to be taken to a staff held on a point of known level. Usually this means that we are looking back and so this is a *Backsight*. In the example given the first backsight is 2.703 to the Bench Mark. The example assumes that we are going to travel forward some distance to get to the area that we want to work in. It is too far to read the staff accurately and so we use two *Change Points* to 'leap-frog' to the area. A change point should be reasonably firm and such that the staff man will be able to put the staff back on the same spot after the instrument has changed position.

Before the instrument is moved to a new position beyond the first change point a reading is taken on to the staff held on the change point. This is a sight looking forward in the general direction of travel, and it is the sight which will carry the chain of levels forward, so it is called a *Foresight*. In our example its value is 2.162 and it is entered in the *Foresight* column. It is written on the second line and in the *Remarks* a note is made that this is a change point.

Now the instrument 'leap-frogs' over the change point and is set up some convenient distance beyond it. The first sight from this new position *must* be a backsight on to the change point that we just used. Its value in our case is 1.285 and it is entered in the *Backsight* column and on the second line down. This line is used because this line refers to the change point, so the foresight of 2.162 and the backsight of 1.285 are both linked to the one change point.

This 'leap-frog' process is carried out once more with a foresight of 1.472 and a backsight of 0.817 to carry us into our working area. In general you can say that the first sight you take after setting up the instrument in a new position is a backsight. The last sight you take before moving the instrument is a foresight. Every change point line in the book should have a foresight entry and a backsight entry.

Now that we have reached the working area we can find the levels of individual points. In the example given levels are taken on to a stone slab labelled A and a grave cover labelled B. The respective staff readings on to these two points are 1.630 and 2.913. Since these are neither backsights, nor foresights they go in the *Intermediate* column with an appropriate entry in the Remarks. Intermediate sights are single sights on to unique objects and so there is only the one entry on each line.

For simplicity our example stops here, but in practice it is important that we level back to the starting point or forward to some other point whose level is known. This provides an excellent check on the work and it is ignored at your peril. Very rarely will you check back exactly

5: Levelling

but you should decide before you start out what you will accept as an allowable error.

So far we have gone through the process of obtaining the readings in the field, but now we must go through the arithmetical process to work out the results. This is called *Reducing* the results. It may be done after the field work is finished or as the work proceeds

The book with levels reduced is shown below.

Back Sight	Intermediate Sight	Fore Sight	Collimation Level	Reduced Level	Remarks
2.703			74.043	71.340	Bench Mark
1.285		2.162	73.166	71.881	Change Point
0.817		1.472	72.511	71.694	Change Point
	1.630			70.881	Stone slab *A*
	2.913			69.598	Grave Cover *B*

Table 5.2 Booking and Reduction of Levels

The method of reducing the book will be better understood by following it through with constant reference to Fig. 5.5.

Under the Reduced Level column on the first line the entry of 71.340 refers to the reduced level of the Bench-Mark. The reduced level means the height above the particular datum used, which in most cases is Ordnance Survey Datum. For the Bench-Mark this value will be available either from previous work or from an OS map.

When the instrument is set up and the first back sight taken on to the Bench-Mark, the *Collimation Level*, i.e. the line of sight, will be *above* the Bench-Mark by a distance equal to the staff reading. Therefore if 2.703 is added on to 71.340 the answer is 74.043 and this is entered in the Collimation Level column, on the first line.

Now when the foresight of 2.162 is taken on to the first change point, this measures how far the change point is *below* the collimation level. Therefore subtract 2.162 from 74.043 to give 71.881. This value is entered in the Reduced Level column against the first change point. This process has meant that we have derived the reduced level of the first change point from the reduced level of the Bench Mark.

Repeat the procedure again for the second change point as follows. Add 1.285 to 71.881 to give the new collimation of 73.166. Subtract 1.472 from 73.166 to give the new reduced level of 71.694.

At the next set up the backsight of 0.817 added to the reduced level of 71.694 gives a new collimation level of 72.511.

Now *note carefully* that as the intermediate sights are taken the level of the instrument is not disturbed. Hence until we come to the *next foresight* the value of the collimation level will stay the same. So for Stone slab A subtract the intermediate reading of 1.630 from the collimation level of 72.511 to give the reduced level of 70.881.

Similarly for Grave Cover B, subtract 2.913 from 72.511 also, to give the reduced level of 69.598.

If the levelling continued with a return to the Bench-Mark the above procedure would carry on until we obtained as the last entry the value for the Bench-Mark. Comparing this with the starting value we would find our error. Even if there is no error check back over the arithmetic very carefully.

On a small site you may be fortunate and be able to command both the bench-mark and all the points you need from one instrument setting. In this case you will only need to take the back-sight and then proceed to read all the intermediates. But do not forget at the end to check back on the bench-mark and record the reading as a foresight, to prove that the instrument has not been disturbed.

5.10 Setting Pegs to a Level

Now you are in a position to find the levels of different objects and it is time to look at the reverse process of setting an object to a given level. For example you may wish to put pegs at the corners of a trench to serve as reference points for measuring down inside the trench and drawing sections. Naturally it will be very convenient if we can make these some nice round numbers. If you turn back to the example of a level book given earlier we can imagine that from the last instrument position we wish to set pegs so that they are exactly 65.000 above datum. This means that when we sight the staff held on such a peg the reading should be such that the reduced level works out to be 65.000. If the staff reading subtracted from the last collimation level of 65.693 gives an answer of 65.000, then equally 65.000 subtracted from 65.693 gives the required staff reading, which is of course 0.693. Set the pegs lightly in the ground and take a sight on them. Now tell your assistant roughly how much the pegs must be driven down. By a process of sighting and

5: Levelling

driving, the pegs can be brought down to the correct level. It is a little easier to make a mark at a given level on a wall or post. In this case the staff can be slid up or down the wall until the correct reading is sighted. Now the bottom of the staff is at the required level. When doing this type of work try to make things easier for your assistant by giving some idea of how much movement you need; such commands as 'up a bit' are infuriating.

There is an alternative way to mark a level on a wall or post. If the staff can rest on a clearly defined point on the ground as it is held against the wall then the level of this point may be found. Now the person on the staff can measure up the wall to give the required level. To do this either the staff or a hand rule may be used.

5.11 Section Datum

When drawing a section it is often useful to mark a level line on it as a datum for the vertical measurements. Nails or arrows are driven in horizontally at each end and a cord drawn taut between them. If the section is very long it may be necessary to put in intermediate points. The staff should rest on the highest point of the bottom of the trench against the section. From the approximate reduced level for this point you can decide what will be a convenient level for the datum line. Once you have decided what value to use then proceed as described above for marking a wall or post.

If you are working on a site which has a steep slope and a section which runs down hill, you may find that a level line which is near the bottom of the trench at the upper end will run out above the surface part way along. In this case you will have to step the datum line down before it runs out. It will make a little more work for you, but the people who are going to draw the section will be very grateful, as such cases can cause them a great deal of trouble.

5.12 Abney Level

When it is required to do levelling work to a lower standard of accuracy than that so far described, or in terrain which is so rough or overgrown as to make such methods impossible, the Abney or hand level may be used.

An Abney Level may be fitted with a low power telescope, but more usually there is none at all. It is a simple hand held device, with an open tube giving a direct sight and some reading mark. A mirror gives a simultaneous view of a small bubble tube and when the bubble straddles the reading mark the sight is horizontal. Sometimes they are fitted with a graduated vertical semi-circle, so that the line of sight can be set to a given slope, or a slope may be measured. The principles of use are exactly the same as for the major types of levels, but of course the sighting range is much more restricted and the accuracy of the level line is not so good.

Although it is a 'hand' level, it is a great mistake to try to use one without some form of support. One good technique is to get a ranging pole and turn it upside down so that the flat end can rest reasonably firmly on the ground. Now hold the Abney level in one hand and while grasping it wrap your fingers around the pole so that the level is held at some well marked point on the pole. This ensures that you keep a reasonably constant height of collimation while taking readings from one spot.

I remember once walking down the side of a small valley which was so densely covered with high bracken that my feet never made contact with the ground, I floated on bracken. A few days afterwards, I watched one of our teams working down the slopes in steps using an Abney level to get the contour. A classic example of using the right instrument for the conditions.

5.13 Clinometer

Some patterns of clinometer form part of the alidade for use on a plane table, but the types to be dealt with here are the small hand-held version. These are small compact instruments, rather reminiscent of a lady's powder compact. They are held in a vertical plane and a sight is taken through a peep-hole which gives a view of objects in the distance. At the same time, just to one side of the view a scale of degrees and/or gradients is visible. This scale is mounted on an inner drum which is weighted so that when released it rotates about its axis and comes to rest with the zero of the scale in the horizontal position. It may be used for similar work to the Abney level such as approximate levelling, measuring or setting out gradients and the like. It should be remembered that the sighting base is really quite small and no optical help is provided; consequently the accuracy is not very great. However for approximate work, or preliminary assessments it can be very useful.

5.14 Boning Rods

Before leaving the subject of levelling mention should be made of boning rods and their use. They are a very useful device for establishing or finding levels from a series of pegs set by a levelling instrument. For example if a trench has level pegs set at its corners any further levelling

5: Levelling

required in or around the trench can be done by boning so the continuous use of a level is not essential. Boning rods consist of three T-shaped pieces of wood. They can be made by anyone with a slight knowledge of carpentry. Two of the rods have stalks of the same length, about 1 metre, and the third may be the same length or longer. The size of timber used can be about 5 cm by 1 cm and the cross bar of the T about 40 cm long. Suppose that we have three boning rods of equal length and that two pegs have been set so that their tops are at the same level. A boning rod is held vertically on top of each peg and the third boning rod is held at some point between the two pegs. An observer at one of the pegs looks along the line and directs the movement of the intermediate rod up or down until the tops of all the T's are seen to be in line. When this is achieved then the bottom of the intermediate boning rod is at the same level as the top of the two pegs. Using this principle a number of possibilities are open. A measurement may be made downwards from the bottom of the intermediate rod to determine the level of some unknown point. A peg may be set at some intermediate point and this used in turn as a fixed point for boning from.

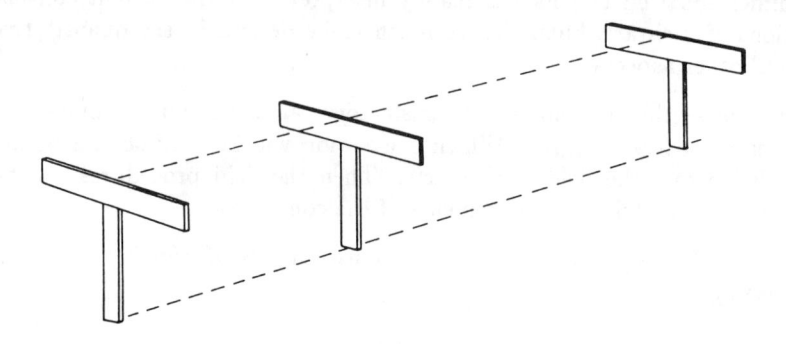

Figure 5.6 Use of boning rods

Alternatively if the two outer pegs are set so as to be at some particular gradient then the intermediate boning rod will fix a point which is on that particular grade line, or if the middle boning rod is longer than the end ones then it will fix a point at so much below the grade line. The same effect may be obtained by holding the middle boning rod against a ranging pole which projects down below the bottom of the boning rod.

Chapter 6

Earthworks

6.1 Plotting Conventions

Hills, valleys and earthworks present problems in their representation. We only have a two dimensional sheet of paper on which to show a three dimensional object. The solution lies in using some convention which is familiar to the person who produces the drawing and to the person who wishes to study it. The success of the convention is measured by the accuracy of the information, both qualitative and quantitative, which is conveyed to the user. Thus an artistic bird's eye view gives a three dimensional effect which is readily interpreted by the current conventions of such art, but it leaves much to be desired in the quantitative and other aspects.

No convention is completely satisfactory; all have shortcomings to a greater or lesser degree. Which convention will be used should be decided before the field work starts. Then the field procedures can be chosen to match the requirements of the convention.

We shall only consider two conventions, namely *Hachuring* and *Contouring*.

6.2 Hachures

This convention is very good in presenting a picture of surface relief. The draughtsman can vary the style of the hachures, their length, spacing and intensity.

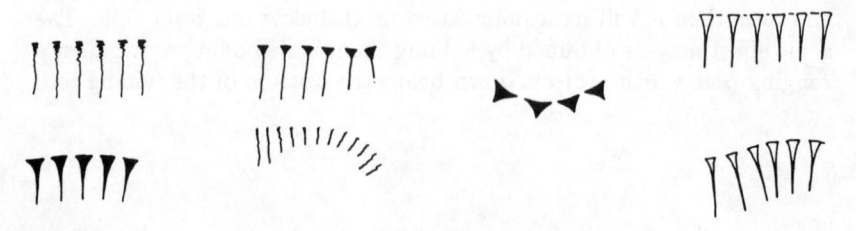

Figure 6.1 Various styles of hachuring

6: Earthworks

Skilfully used, subtle variations in gradient and differences in height may be brought out. Sometimes hachures are drawn with a stiff, mechanical precision. This gives a neat and tidy appearance, with an air of exactness. Unfortunately real slopes are rarely akin to a row of soldiers sized off on parade.

The main failing of hachures is that they do not give any indication of the numerical differences in level. They show high spots and low spots, and to some extent which are higher high spots and lower low spots, but they do not reveal by how much they are higher or lower. What hachures do indicate are the areas of well defined slopes and the relative gradients involved.

If it is known that the plot of a survey will make use of hachuring then the field operations will be directed towards finding the lines which separate areas of different gradient, such as ridge lines, valley lines and lines of change of slope.

6.3 Contours

Contours have the big advantage that they allow a numerical value to be put upon differences of level. However they are not always as easy to interpret as hachures. In addition contours may not show up minor features unless the contour interval is quite small, but small contour intervals bring in their train a number of other problems. When assessing the picture presented by a contour map the first requirement is to know which is up and which is down. This may sound very obvious, and so it is when using a topographical map covering a reasonable area. In such cases the presence of a coast line or rivers and streams will give an immediate indication of surface forms which almost subconsciously is used to translate the picture presented by the contours.

All contour plans should have some of the lines labelled with their values. This is particularly necessary with large scale plans of small areas. Some archaeological reports show contouring at the bottom of building foundation trenches with no labels attached. Such 'contours' are not meaningless; they can be given any meaning you care to put upon them! Fig. 6.2 shows a contour plan where without further information none of the possible sections illustrated can be proved to be wrong - or right.

The difficulties of contour interpretation may be illustrated by an example (Fig. 6.3). Try to imagine a ditch and double bank earthwork which is roughly rectangular in plan. Part of one end has almost been obliterated and erosion processes have created irregular humps in the banks and ditches. Finally place the whole complex on a piece of ground

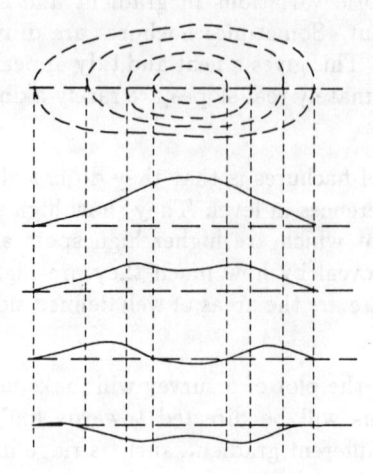

Figure 6.2 Contour ambiguities

which has a marked slope roughly in the direction of one of the diagonals. The resulting contour pattern will take a lot of unravelling by someone who has never seen the site, but hachures tell the story fairly quickly. The best solution is to make two drawings, one of hachures and the other of contours. The hachure sheet will give a visual impression and the contours record the physical dimensions.

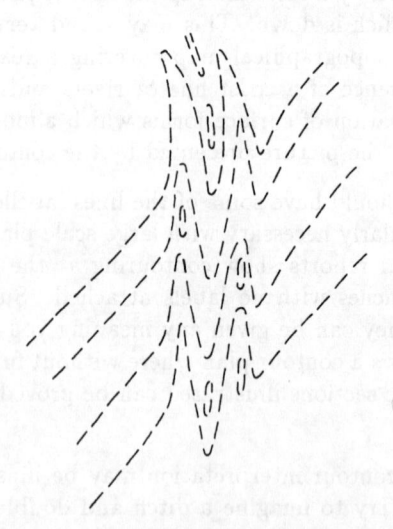

Figure 6.3 Contours for a complex earthwork

6: Earthworks

6.4 Types of Contouring

Surveying operations for contouring can be divided into two main types. In one of these, points on the ground which lie on a given contour are found and then their position is located by some appropriate surveying technique. The other method is to find the level of a number of points over the area and then to interpolate the contours between them. In the latter case the points may follow a regular pattern, such as grid intersection points or they may be irregular in plan but chosen with an eye to the topography of the area.

6.5 Contour Chasing

The first of these methods is often called 'chasing the contour' since the staffman moves along the line of the contour trying to find its position (Fig. 6.4).

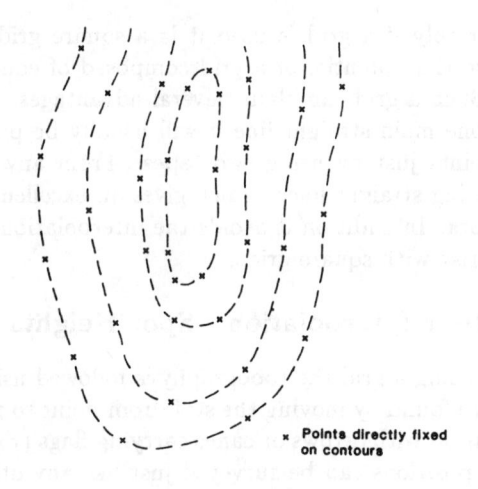

× Points directly fixed on contours

Figure 6.4 Contour chasing

To achieve this a level is brought into the area and its height of collimation found. Suppose this to be 67.314. If the 65.000 contour is required then the staff reading should be 67.314 − 65.000 = 2.314. The staffman moves about, trying different points, and moving up or down as required by the person on the level until the required reading, or a value sufficiently close to it is obtained. This position may be marked with an arrow or a cane with a flag on it. Subsequently, or simultaneously, its position is found. For example a series of lines forming a framework may have been laid down beforehand, and the point is surveyed relative to one of these lines as described in the section on detail survey. An alternative method might be to have a theodolite set up at some adjacent survey station and find the position of each contour point by tacheometry (See Chapter 8).

6.6 Contour Interpolation - Grids

The decision to use a grid should only be taken after careful consideration of the nature of the ground and the size of mesh of the grid. For example consider a gently sloping area traversed by a meandering stream running in a narrow gulley. If a large mesh grid is put down it might miss the stream altogether, or perhaps only get one or two points down in the bed of the stream. From the evidence of the levels on the grid a completely wrong picture would emerge. The mesh must be fine enough to catch the detail. In the case quoted above a fine mesh would be wasted over much of the area.

Almost invariably if a grid is used it is a square grid, and not many people realise the potential of a grid composed of equilateral triangles (Fig. 6.5). Such a grid can claim several advantages. First of all after setting out one main straight line it will usually be possible to set out the other points just by using two tapes. From any one point there are six radiating straight lines, which gives an excellent opportunity to verify the work. In addition it avoids the interpolation problems which sometimes arise with square grids.

6.7 Contour Interpolation - Spot Heights

If instead of using a grid the topography is followed using spot heights, then levels are found by moving the staff from point to point, and marking each position with arrows or canes carrying flags (Fig. 6.6). Then as before these positions can be surveyed just like any other detail, using the technique which best fits the situation. The use of tacheometry is very suitable for this technique because the level of the point and its plan position are both recorded in the one operation.

6.8 Assistants

In all surveying operations experienced assistants make a great deal of difference to the speed and smoothness of the work, but such help is particularly important in contouring. To follow a contour or to choose good spot heights to bring out the topography requires skill, intelligence and no little experience. Ideally the most experienced person should act as staffman, but as he is usually required to operate the instrument the staff is given to someone else. With a person of little experience on the

6: Earthworks

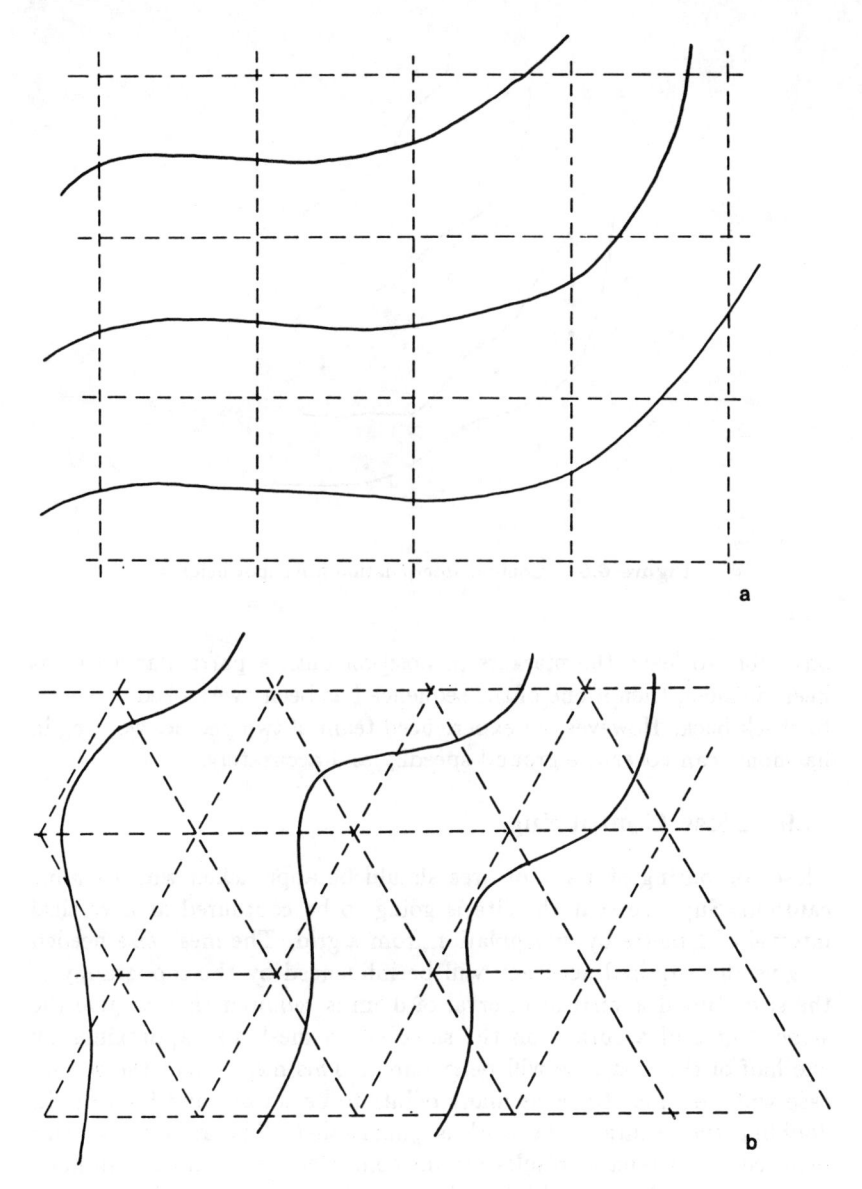

Figure 6.5 Rectangular and triangular grids

staff it may be necessary to escort him round the area and pick out the staff positions for him, marking them in some way. This is a slow process, very wasteful of time and hence costly but the only sure way of getting a true interpretation of the land. In such a case it is usually

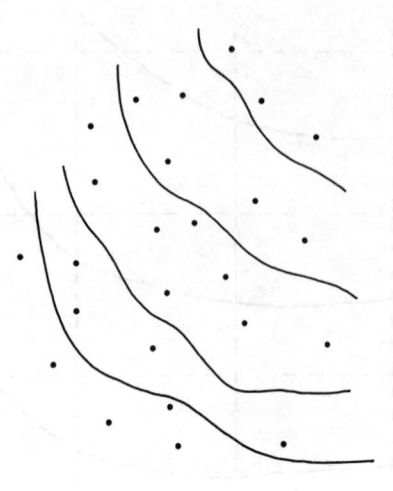

Figure 6.6   Contour interpolation from spot heights

advisable to leave the markers in position until a particular area has been finished, then if one of the sequence has been overlooked it is easy to check back. However, an experienced team of two people, working in harmony, can cover the ground speedily and accurately.

6.9 Close Contouring

Close contouring of a site or area should be approached with extreme caution. Suppose that the site is going to be contoured at a vertical interval of 1 metre by interpolation from a grid. The mesh size needed to give the required accuracy will be influenced by the topography of the site. But if a vertical interval of 0.5m is required then to give the same degree of accuracy on the same site a mesh size approximately one half of the first case will be required. This means that the second case will need four times as many points to be set out and levelled. In doubling the accuracy the work is quadrupled. The amount of work required to get reliable results for fine contouring is far more than most people realise. As vertical intervals get very small other problems arise. Natural irregularities in the ground become a significant proportion of the quantity being measured. A tussock of grass can give a difference of level of 7-10cm.; a fresh molehill can be about 22cm. high. No one will mistake a fresh molehill, but what will it be like after about six months? Its height will be reduced a little but now disguised by vegetation. A very small mesh grid is almost certain to have a significant number of

6: Earthworks

points fall on natural irregularities. Incidentally this makes the point that when chasing contours you do not need to be too persistent in getting the correct value; the horizontal movement required will probably not show up on the plot. Besides it is not unknown for an overtried staffman to provide a slightly higher level by resting the staff on the toe of his boot!

Chapter 7

Theodolites - Basic Usage

7.1 Types of Theodolite

Theodolites may be divided into different categories by the way in which the angular scales are read. Since their first practical development in the sixteenth century there have been a number of ways of doing this, but now there are only three types which you are likely to meet. The largest group uses optical means to achieve precision in the readings, usually in the form of an optical micrometer. The type which is old-fashioned and now practically obsolete is the vernier theodolite. The latest type and the one which, although very expensive, is becoming more common is the electronic theodolite. In this type of instrument the graduation of the scales is not in a form which is intelligible to human beings. They are scanned electronically and the results presented as a digital display.

It would be wasteful to give a description of how to read a vernier because it is unlikely that you will ever need the knowledge. Verniers are still used in some accurate linear measuring devices. If you are familiar with these, and if you are called upon to use a vernier theodolite you should be able to work out the angular system for yourself. To help you to do this it is useful to know that most theodolite verniers measure either to 20 or 30 seconds. If you want to know more about verniers then look in any book on surveying which was first published before 1980. Such books usually have a section describing the vernier in detail.

In what follows we shall talk of electronic theodolites or optical theodolites. The latter term of course refers to the means of reading the scales: all theodolites have an optical system in the telescope. If the matter being discussed applies to both these types then we shall use the word theodolite with no distinction as to type. Anything which applies to optical theodolites will usually apply to the vernier instruments.

7.2 Handling

At first sight a theodolite seems a bewildering array of wheels, knobs and screws, specially designed to lead the beginner into doing the wrong thing. The way into the puzzle is to think about the basic things which are required of the instrument and this we shall look into shortly. When handling theodolites be firm, gentle and smooth. Jerky movements will

7: Theodolites - Basic Usage

tend to disturb the instrument setting and if you find yourself using force you are almost certainly doing the wrong thing.

The theodolite measures horizontal angles and vertical angles. If it has not been mistreated it can do this very accurately. At first you will be mainly interested in the horizontal angles, the angles as seen on a plan, and you will not need to bother about the vertical angles. There are two graduated circles on a theodolite. One is called the horizontal circle and is used to measure the horizontal angles. The other is called the vertical circle because it is used to measure the vertical angles.

The horizontal circles of optical theodolites are nearly always graduated clockwise, so starting from zero the readings increase as we turn to the right until as we turn full circle and back to the starting point the reading becomes 360°. To turn off a right angle to the right the reading changes from 0° to 90°; to turn off a right angle to the left the change is from 360° to 270°. When the reading is 180° the instrument is pointing directly opposite to the 0° position. A few optical theodolites have a system which allows you to choose between having the graduations increase in a clockwise or anti-clockwise direction. Electronic theodolites usually provide this choice of rotation at the press of a key. This is not the advantage that it may seem. You will need to record very carefully which mode you are in if you are to avoid confusion when handling the results. It will probably be much safer to work always in the clockwise direction. Then no matter what instrument you use you will know what the values mean.

The simplest geometrical figure containing angles is a triangle and this is the building block for many surveying networks. It has the advantage of providing a simple check on the readings since the sum of the three angles should be exactly 180°. In practice when reading the angles to the accuracy provided by a theodolite, it will be found that small errors are bound to creep in. However, as long as the total is sufficiently close to 180° the readings may be accepted, but any unacceptable errors will immediately be obvious. Try to keep in mind the scale of things that you are dealing with. There are 360° in the full circle, with 60 minutes to each degree and 60 seconds to each minute. Theodolite angle scales can commonly be read to at least 10 or 20 seconds. An angle of 20 seconds is about equal to the diameter of a 10p coin held at a distance of 290 metres.

7.3 Plumb-bob, Optical Plummet

To centre the instrument over a survey station either a plumb-bob, an optical plummet or a plumbing-rod will be provided. The plumb-bob is simple and direct, but it becomes difficult to use in a strong wind. However it can be used quite successfully if shielded and used with care. Usually provision is made on instruments fitted with an optical plummet or a plumbing rod for a plumb-bob to be used if desired. An instrument fitted with an optical plummet has a small eye-piece somewhere in the lower part of the body. A right angle prism gives a sight vertically downwards, so that the theodolite can be set directly above the station. But always remember that the optical plummet is not truly vertical until the instrument has been levelled. For this reason it is best to do the initial work using the plumb-bob. It is pointless trying to level the instrument before it has been approximately centred. Very often this type of instrument has a small circular bubble closely associated with the optical plummet, and if this is levelled the optical plummet will be vertical, or sufficiently near for centring.

7.4 Plumbing Rod

The plumbing rod is a special feature of all Kern theodolites. It is a telescopic, central rod which at first sight seems to be an extra leg to the tripod. The pointed end is placed at the centre of the station marker, and the tripod top is manœuvred until the plumbing rod is vertical. This is indicated by a small circular bubble carried on a bracket projecting from the side of the plumbing rod.

7.5 Setting Up - Centring

To set up a theodolite place it so that it is approximately central above the station with the top of the tripod nearly level and the feet of the tripod just resting on the ground. Take time and do not rush this stage. When you are satisfied with the initial position, press the legs firmly into the ground. Usually they will all go in about the same amount, so keeping the instrument approximately centred and approximately levelled. If one leg goes in more or less than the other two then the length of the leg can be adjusted to compensate.

Next by using the adjustment on one or more of the legs, centre the instrument more exactly. Even if the theodolite is fitted with an optical plummet it is best to use the plumb-bob up to this stage. Now the final precise centring can be done using the movable head adjustment on the instrument, or between the instrument and tripod top. The optical plummet, if fitted, can now be brought into play and this will call for an approximate levelling of the instrument, using the circular bubble.

7: Theodolites - Basic Usage

7.6 Setting Up - Levelling

Once the instrument is properly centred the next stage is to level it using the plate bubble. Sometimes a bubble tube is fitted high up on one of the standards supporting the axis about which the telescope tilts. This is NOT the plate bubble. The one to use will be found somewhere on the main body of the lower part of the instrument. The procedure for levelling is as described in the section on levelling the dumpy level. A shortened version is given here.

Centre the bubble in its tube when it lies parallel to two foot screws (position 1). Do this by rotating the two foot screws equally but in opposite directions. Turn the instrument through 90° so that the bubble lies parallel to the direction of the third foot screw (position 2). Now centre the bubble using the third foot screw only. Return to position 1, re-centre if necessary. Do the same in position 2. When no further adjustments are necessary in positions 1 and 2, rotate the instrument through 180° from position 1 and call this postion 3. If the bubble moves from the central position by an appreciable amount, say about 1 division or more, bring the bubble back *half way* using the two foot screws. Now turn the instrument through a further 90° so that it is in position 4, 180° from position 2. Using the single footscrew bring the bubble to the same *half way* point. Work between positions 3 and 4 until the bubble settles at the same *half way* point in each case. Now check back to position 1. If there is any further appreciable displacement of the bubble from the *half way* position split the difference and treat this as the new *half way* position. Keep on with the process until the bubble maintains a constant position (or nearly so) whatever the position of the instrument. In practice the first reversal will often suffice and the levelling is quickly completed.

Now do a quick check that the centring is still satisfactory.

7.7 Setting Horizontal Angles

When the theodolite has been properly centred and levelled we can turn our attention to measuring, or setting out, angles. It is impossible in a short text like this to cover all the variations in methods of reading the scales of angles provided by different instrument makers. All that we can do is to look at broad general types and leave the reader to apply these to particular instruments that he meets. It is strongly recommended that you have a theodolite in front of you while you read this section.

In optical theodolites the scales of angles are carried on glass circles. If an angle or a number of angles are to be measured from a station one of the sight lines is chosen as a starting point. When the telescope is set along this line it is very necessary that the observer should be able to set the circle reading to whatever value he chooses. For example he may elect to make that particular reading 0°0'0". The way in which this is done divides theodolites into two main groups. In one of these groups the observer can rotate the glass circle inside the instrument and so set it to some chosen value. This is done by rotating a knob at the side of the instrument body. This knob is protected by a hinged cover to prevent accidental displacement of the circle. Instruments of this type have only one clamping screw and a corresponding slow motion screw to control the rotation of the instrument for measuring horizontal angles. Electronic theodolites work on this same principle but the circle which carries their graduations is not physically rotated. The necessary adjustment is done electronically by pressing one of the keys.

Theodolites which do not have provision for turning the glass circle, as described above, are fitted with two sets of clamping screws and slow motion screws, one set above the horizontal circle and one set below it. Vernier instruments have the same arrangement. If the lower clamping screw is shut the circle is locked to the bottom part of the instrument. Now if the upper clamp is open the telescope may be turned horizontally until the horizontal scale is approximately at the chosen value, whereupon the upper clamp is closed.

Next using the upper slow motion screw the scale can be set to the exact value required. At this stage both clamps are in the closed position and so the whole instrument is held rigidly to the base. Now release the lower clamp and swing the telescope horizontally to sight the initial point. In doing this the circle will be carried round locked to the upper part of the instrument and so keeping the reading at the chosen value. Align the telescope approximately onto the point and close the lower clamp. Use the lower slow motion to bring the telescope into exact alignment. The instrument is now directed towards the starting point and the horizontal circle reads the chosen value. To proceed to measure the angles release the upper clamp and swing on to the various points, at each one closing the upper clamp and using the upper slow motion to get the correct alignment.

7.8 Reading Angle Scales

The glass circles on an optical theodolite are viewed through a special reading eyepiece, situated either adjacent to the telescope eyepiece or somewhere on one of the side supports. Illumination is provided by a mirror which is hinged and which can be rotated so as to get the most effective lighting on the scales.

7: Theodolites - Basic Usage

The scale for reading horizontal angles may be marked either as H, or Az standing for Azimuth. The vertical angle scale will be marked V. Both horizontal and vertical scales may be seen at the same time. If not then somewhere on one side support there will be a knob which when turned will change over the scales from H to V or vice versa. The knob will carry corresponding markings which will identify its function.

The scales may be clearly set out to give a direct reading of degrees, minutes and seconds. A brief study of the scales, noting how many subdivisions go to make up a degree, will usually reveal what values the sub- divisions represent. Such an instrument will give a reading to about 20″. Those instruments which can be read to a smaller value than 20″ will usually incorporate some form of optical micrometer. When looking through the reading eyepiece a separate subsidiary scale will be seen, in addition to the horizontal and/or vertical scales. The micrometer control knob will be found on one of the side supports. Turning this knob will make the main scales move across the field of view, and there will be an index mark, or other feature to align on. The nearest division on the main scale is brought into coincidence with whatever mark is provided. As this is done the subsidiary scale will record the movement in minutes and seconds. Some instruments will give a reading to within 5″, others may be read to 1″. The way in which the scales are divided up varies from instrument to instrument, but if the micrometer head is run from one end of its travel to the other its complete range can be found. If this is then compared with the movement and divisions of the main scales it is quite easy to work out the graduation of the instrument.

When setting an optical micrometer instrument to a given reading and aligning it on a chosen starting point, the minutes and seconds MUST be set to the desired values right at the beginning. Suppose that a reading of 0°0′0″ is required and that the main scale is set to 0° before the optical micrometer is set to 0′0″. Then when the optical micrometer is used to give a reading of 0′0″ the main scale will move off the 0° position. A simple trial on an instrument will soon demonstrate the correct sequence to follow.

If an electronic theodolite is being used the required readings are displayed by pressing the appropriate keys. Different makers have different arrangements, but usually there is provision for reading the horizontal and vertical angles and for setting the horizontal angle to whatever starting value is required by the observer when sighting the initial station.

7.9 Vertical Angles

Theodolites vary in the position of the zero reading on their vertical circles. In some cases they read zero when the telescope is pointing vertically upwards, that is towards the zenith. Other instruments are made with the zero reading pointing vertically downwards, that is towards the nadir. In both these cases the graduations on the circle go from 0° through 90°, 180° and 270° to 360°, which of course is marked as 0°. To find out how a particular instrument is graduated set the telescope approximately level. Now look at the vertical circle reading and at the same time gently tilt the telescope upwards. It is a good idea to put the details on a label inside the box-lid so that you can always refer to it later.

There are other patterns of graduation of the vertical circle which you may meet, particularly on older instruments. A quick examination as described above will soon tell you how the readings run. It is a wise precaution to check the reading with the telescope approximately level and pointing in one direction, then turn it right over to check in the reverse direction. You may find that it reads zero for the horizontal in both directions.

The procedure for levelling a theodolite given in Section 7.5 above, mentioned a bubble-tube which might be found high up on one of the standards supporting the horizontal axis. This bubble-tube, which is more sensitive than the plate bubble, controls the setting of the vertical circle index or reading mark. Lower down on the same standard there will be a knob which is used to adjust the setting. Turning this knob allows you to bring the bubble to the centre of its run. It is only when this is done that the reading on the vertical circle gives the true angle of the telescope to the horizontal. This setting may be disturbed as the instrument is turned about the vertical axis, so it should be checked for each new point observed.

Some modern theodolites do not have the upper bubble as just described, instead they are fitted with a vertical circle index compensator. The compensator automatically keeps the index mark in the correct position, once the instrument has been levelled using the plate bubble. The action is similar to that of an automatic level but the internal mechanism is different. This type of theodolite is very useful for any surveying which involves taking a lot of vertical angles, as for example in Tacheometry (see Chapter 8).

Chapter 8

Tacheometry

8.1 Basis of Method

Tacheometry is a method of surveying in which the theodolite is used not only to measure angles but to measure distances. To do this a staff is used as an accessory to the theodolite. Usually a normal levelling staff will be very suitable for the purpose. The method is economical in man-power and the fieldwork can be completed quickly, but it should be used with discretion since it may not always offer the best solution.

When using a theodolite or level it will usually be apparent that as well as the main cross wires running horizontally and vertically right across the telescope field of view, there are two short horizontal lines equally spaced above and below the main horizontal (Fig. 8.1). These are the stadia lines and they enable the instrument to be used as a tacheometer (sometimes they lead to a mis-reading, especially when levelling).

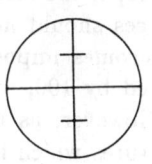

Figure 8.1 Typical stadia lines

The simplest application of tacheometry is when using a level, but it is not as flexible as when a theodolite is used. However it forms a good introduction to the principles involved (Fig. 8.2). When a staff is sighted through a level the line of sight is horizontal and the staff is vertical. The staff is at right angles to the line of sight and under these conditions

$$D = 100(SH - SL)$$

where D is Distance, SH is the Highest Staff Reading, and SL is the Lowest Staff Reading.

If the staff readings are 2.173 and 2.0l8, then

$$D = 100 \times (2.173 - 2.018) = 15.5$$

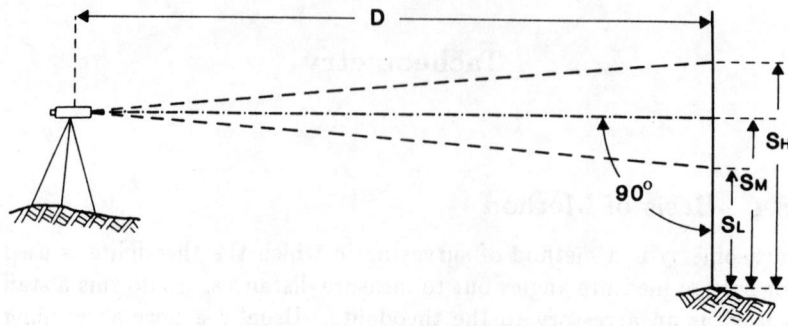

Figure 8.2 Using level as tacheometer

The value of the multiplying constant, 100, depends on the optical design of the instrument; only very rarely is any other value used.

There are two things to note here. The first is that the units for D are the same as the units in which the staff is marked. A metric staff gives the result in metric and an Imperial staff will give an Imperial result. The second thing to note is that the accuracy of the result depends upon the accuracy with which the staff is read. This in turn depends on the power of the telescope, the distance of the staff and the lighting conditions. Sight distances should not exceed 175m as at this range accurate staff reading becomes impossible. Any inaccuracies in staff reading are also multiplied by 100, so that the measurements are not in the precision class. However, as long as the method is used with discretion in locating features which by their nature are not precise, it will be found to be of great power and flexibility. Of course it may also be used when the scale of the final plot is so small that the inaccuracies of measurement are below the threshold of plotting accuracy. If contours or spot heights are required they may be obtained with little extra work. A good standard procedure is to take all readings as though levels are required. It is almost as quick to do this as it is to remember to omit part of the routine, and sometimes the extra information comes in useful.

8.2 Inclined Sights

The method as described so far has been for a horizontal line of sight, but this of course is very restrictive. If a level fitted with a horizontal circle is used to measure an angle or set out a right angle it can often prove to be a frustrating experience. The critical points of observation often lie above or below the telescope's field of view. With a theodolite the ability to tilt up or down greatly increases the area which may be covered. However, as with all benefits, a price has to be paid, and in

8: Tacheometry

tacheometry the price is an increase in the calculations to be done.

Although the line of sight may now be inclined to the horizontal the staff is held vertically, and corrections are made by including the angle of tilt of the telescope in the calculations.

8.3 Formulæ

Before the era of the electronic calculator and computer tacheometric calculations, although simple, were tedious to do, even with tables to help the work along. Now the position is different and the calculations can be done at a speed to match that of the fieldwork. Not only that, but the work reduces to a simple routine. When using a calculator only a little practice is required in pressing the buttons as specified by the calculator instruction booklet. If a computer is available the task is easier still. Most archæologists or other field workers have some connection with a University or a similar institution, where there will be a computer. A competent programmer will quickly and easily write a suitable program.

Going on now to the formulæ used for an inclined sight, let the higher, lower and middle staff readings be SH, SL and SM respectively as shown in Fig. 8.3

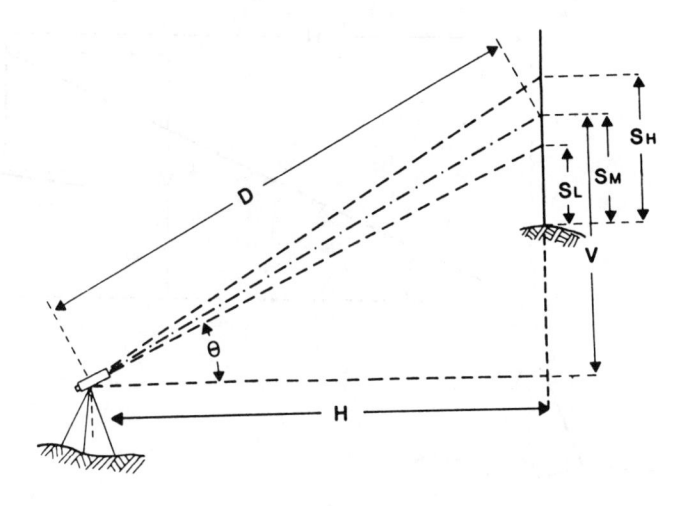

Figure 8.3 Using theodolite as a tacheometer

Let the angle of inclination of the telescope, up or down from the horizontal be θ, then

$$\text{sight distance,} \quad D = 100(SH - SL)\cos\theta$$

The horizontal distance H from the vertical through the theodolite to the staff is given by

$$H = 100(SH - SL)\cos^2\theta$$

The vertical distance V from the horizontal through the theodolite to the reading of the middle line on the staff is given by

$$V = 100(SH - SL)\cos\theta\sin\theta$$

If the instrument is set as shown in Fig. 8.4 at a height h above a station of reduced level Ls, then the reduced level of the point at which the staff is held Lp, is given by

$$Lp = Ls + h \pm V - SM$$

The $\pm$ sign in front of V arises because the sight may be inclined upwards or downwards.

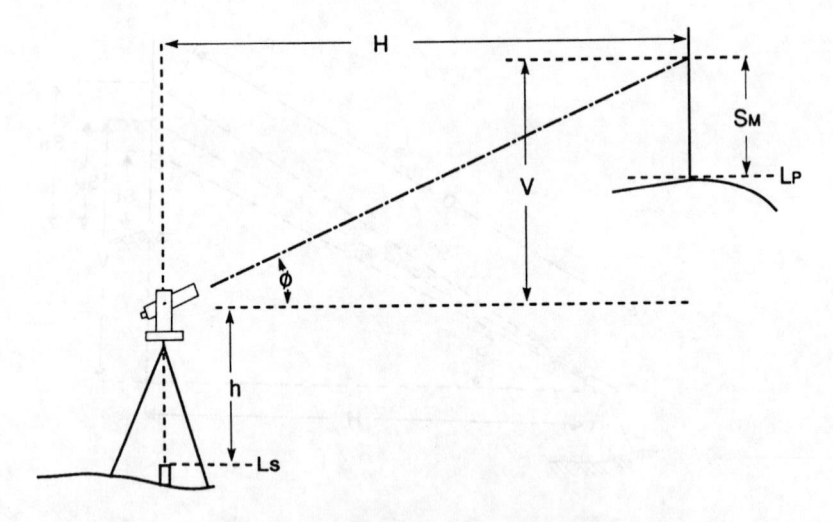

Figure 8.4 Finding level by tacheometer

8: Tacheometry 71

8.4 Application of method

Once the instrument and calculating techniques are understood they can be put to use within a general surveying framework. Tacheometry is part of the tactics of surveying, but the overall framework in which it is used is the strategy, and both are important in producing the final result.

The positions of the survey stations from which the tacheometer will be used may be fixed by any surveying method which will give the required accuracy. The absolute position of each point must be known within an error which will not be apparent on a plot at the largest likely scale to be used. The stations must be so distributed over the area that no tacheometric measurement will contain an error that will be apparent on the plot. Usually if the points are carefully chosen to command an area of detail without using long sights it will be found that the work proceeds faster anyway.

One way to fix the stations is by means of a triangulation system over the site. All the angles in each triangle should be measured (Fig. 8.5).

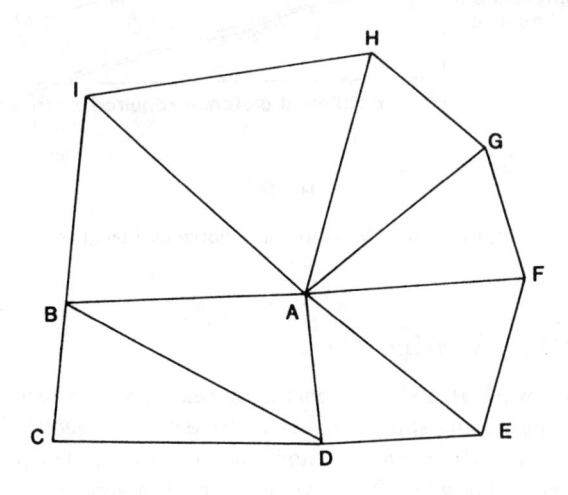

Figure 8.5   Triangulation system used at Barhill Roman Fort

In addition read the angles to any other stations which are visible. This takes only a little extra time and it can be useful for checking purposes. One of the lines which is of a reasonable length and level, or nearly so, is measured accurately with a steel tape. This is done at least twice or three times and accepted if there is agreement to within 3-5mm. The level of each end is found and if necessary the length is corrected to the horizontal (Fig. 8.6). This is done by using the formula

$$H = \sqrt{M^2 - d^2}$$

where H is the Horizontal distance, M is the measured length and d is the difference in level. If none of the sides of the triangulation system are suitable for measurement then a line must be laid out specially and connected into the system by triangulation. From the data the positions of all the stations can be fixed. All this work is well within the capacity of two people and it can be done quickly by an experienced team. It is important to walk all over the site at the beginning, two or three times, putting in ranging poles in tentative positions, then they can be moved and adjusted to give the best layout. Care and thought at this stage pay dividends later in simplifying and speeding up operations.

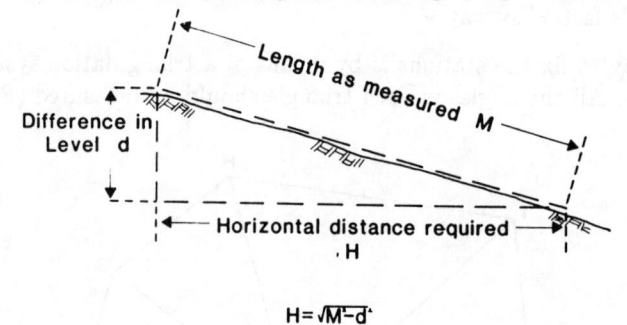

Figure 8.6 Measured and horizontal lengths

8.5 Station Routine

Tacheometric work at a station begins by taking a zero reading for the horizontal angles using some convenient Reference Object. The R.O. is usually one of the other main stations, not necessarily the nearest one. If the R.O. is a long way off the angular error in sighting to it will be small. Then the readings are taken on the points of detail which are to be surveyed. This is done to an agreed plan and based on a good sketch of the area around the station. As well as taking the stadia line readings on the staff, and the vertical angle it is necessary to take the horizontal angle reading. This will give the direction to the point, which combined with the horizontal distance H fixes its position. When taking the vertical angle remember to level the vertical circle index bubble if one is fitted. The points should be numbered consecutively over the whole site. In this way a point number can mean only one unique spot.

When all the tacheometer readings have been taken, the R.O. should be sighted and the zero angle checked. If there are a lot of points to observe from one station it is prudent to check the zero reading periodically then should an error be discovered the number of readings thrown into doubt is limited. In a day's work up to a hundred sets of readings may be taken so it is important that the booking be done neatly and in an orderly manner, supported by a good, detailed, anotated sketch.

Instrument – Tavistock. Imperial* Staff. Station III·3 Zero on Ref. Obj. IV·2. Reading 0° 0′ 0″
Height of Instrument above Station 1·045m

Point	Staff Low	Staff Mid	Staff High	Horiz. Ang ° ′ ″	Vert. Ang ° ′ ″	Remarks
26	7·40	7·46	7·52	179 22 54	268 35 44	↑
27	8·00	8·14	8·27	118 40 44	264 53 38	
28	8·00	8·28	8·57	123 14 58	264 58 23	Inner Bottom
29	7·00	7·45	7·90	136 46 57	266 7 8	of Bank
30	6·00	6·59	7·18	146 29 32	266 47 3	
31	6·00	6·49	6·97	160 18 23	267 13 26	
32	6·00	6·33	6·65	173 47 44	267 5 16	
33	6·00	6·25	6·50	193 11 55	267 45 10	↓
34	9·00	9·15	9·29	170 29 16	268 15 29	Middle
35	8·00	8·30	8·61	153 28 32	266 35 45	Middle
36	5·10	5·19	5·28	321 34 36	269 58 36	↑
37	9·00	9·16	9·32	54 25 30	270 4 58	
38	8·00	8·26	8·53	94 54 47	266 5 50	
39	8·00	8·40	8·80	108 6 30	265 39 42	Outer Bottom
40	6·00	6·49	6·98	125 0 53	265 43 58	of Bank
41	8·00	8·56	9·11	131 57 27	267 17 10	
42	8·00	8·67	9·33	138 44 33	267 24 27	↓
43	7·00	7·76	8·53	143 56 0	266 56 8	Tip of Tongue
44	7·00	7·83	8·67	143 21 0	266 50 0	Ditch Bottom
45	6·00	6·56	7·12	168 14 43	267 39 55	" "
46	4·00	4·39	4·79	207 29 15	268 58 16	" "
47	7·00	7·95	8·89	142 56 25	268 34 13	Road side
48	7·00	7·82	8·65	151 10 43	268 34 24	" "
49	6·00	6·62	7·24	170 42 4	269 13 44	" "

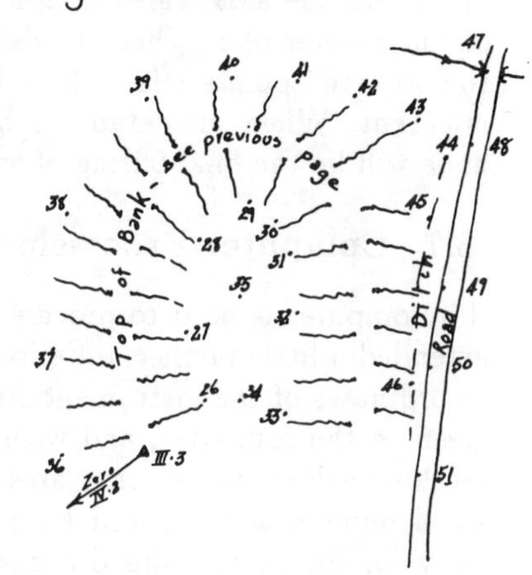

* Converted to metric in computer programme (when necessary)

Figure 8.7 Tacheometric survey readings

8.6 Plotting

It is possible to get a computer plot in which each point is drawn in its correct grid position with its serial number or reduced level written alongside. This will be discussed later; for the time being the manual approach to plotting will be described. For each set of readings the horizontal distance H can be calculated as described earlier, and if required, the reduced level of the point, Lp, may be found. The horizontal angle between the R.O. and the point, along with the distance H, fixes the position of the point. For purposes of plotting it is convenient to treat the line from the station S to the R.O. as one axis of a co-ordinate system and calculate the X and Y co-ordinates. When doing this remember that theodolites are nearly always graduated to give clockwise angles. To get the angle required on the plot, subtract the recorded angle from 360°. So for each instrument station the plot of the points surrounding that station can be drawn out on a sheet of paper.

Now on a sheet of tracing paper or drafting film, a plot of the overall surveying framework is made, giving the position of each instrument station in its correct relative position. The way in which this is done will depend to a certain extent on the method chosen for establishing the framework on the ground. Each station plot in turn is positioned below the overall plot, so that the instrument station is in its right place and the axis which points to the R.O. points to the right spot on the master plot. Then the detail can be drawn in corresponding to the observed points. Here the value of good field sketches will become apparent. When the detail has been copied all that will remain to be done will be the final fairing of the plot.

8.7 Computer Processing

If a computer is used to process the results, then the method may be extended a little further, if desired. Information concerning the overall co-ordinates of the instrument station and the bearing of the R.O. are given to the computer, and with a small addition to the program the machine will output co-ordinates of each point observed, along with its serial number and reduced level. Plotting can then be done straight away on the overall site drawing. When a graph plotting facility is available as part of the computer system then the complete plot of the points can be made by the computer. The serial number or reduced level of each point may be printed alongside the plotted position. It is useful to have a plot of both because if any errors arise it is easier to check back through the serial number as an identifier.

When a large number of points are to be surveyed it is well worth while booking the results on to a computer coding sheet directly. This will save tedious copying from the field book. If the input format is arranged to cater for this it is possible to book on to the left hand side of a coding sheet, leaving the right hand side free for the sketches. This can then be used for entry into the computer without the need for copying, which helps eliminate one source of error and saves a great deal of time. It is thus possible for the surveyor to take the readings in the field and he has nothing further to do until he receives the plot of the points back from the computer.

It is possible to use a small hand held computer to record the readings electronically in the field. It will still be necessary to make good detailed sketches, but the speed and reliability of the transfer to the main computer will be greatly improved. There are a number of these small

8: Tacheometry

instruments available and rapid developments are taking place in this area of technology. Before deciding on a particular model the whole field should be investigated. The characteristics and limitations of the different types should be considered. Needless to say the most important requirement is compatibility with the main computer. In this respect make sure that you know all the accessories and software that will be required to transfer the data reliably and quickly.

Chapter 9

Surveying Calculations

9.1 Introduction

For most archæological surveying the techniques described so far in this booklet will be quite adequate. However, the power and efficiency of surveying operations can be considerably extended by a slight acquaintance with elementary trigonometry. What used to be difficult and laborious operations can now be done quickly and easily, since with a small electronic calculator the trigonometry becomes easy to handle. This chapter shows you how.

Most people are familiar with the use of simple calculators for the basic arithmetic operations of addition, subtraction, multiplication and division. Calculators which will also handle trigonometry cost very little more and are equally simple to operate. Usually it is only necessary to key in the value of an angle, followed by the name of a trigonometric function and the appropriate value of the function will be displayed and may be used in the calculations.

One difficulty that you might meet lies in the system of recording angular measurements. Almost invariably in British surveying, angles are measured in degrees, minutes and seconds. There are two other systems of measuring angles and calculators usually have a little switch so that any of the three systems may be used. This is easier and less expensive than having to use more than one calculator. The units for the other two systems are called Grads and Radians. When you are using your calculator for trigonometrical work make sure that it is switched on to degrees. The three way switch will probably be labelled Deg. Grad. Rad.

On the Continent the commonly used system is based on Grads. In this system a right angle measures 100 grads exactly, and in a full circle there are 400 grads. The grad is subdivided decimally, so that 60 degrees becomes 66.6667 grads. With growing European links you may find it useful to remember this system. There is a third system, mainly used by scientists and sometimes by engineers, in which the unit is a radian. In this system 180 degrees becomes 3.14159 radians. You may recognise that number but you do not need to know why that value is used. You may be interested to know that 60 degrees is equal to 1.047198 radians

9: Surveying Calculations   77

which gives you some idea of the relative values of angles measured in the three systems.

9.2 Sine, Cos and Tan

In the right angled triangle ABC shown in Fig. 9.1 the ratio of the lengths of the sides BC and AB is called the *sine* of the angle A. Now sine of the angle A is usually abbreviated to $\sin A$ and so we can write

$$\sin A = \frac{a}{c}$$

sin A = a/c

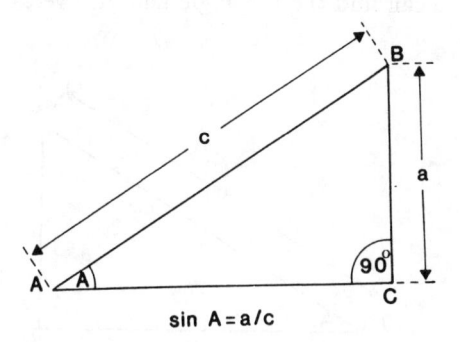

Figure 9.1 Right angled triangle, showing $\sin A$

If we make the triangle bigger, but keep all the angles the same value, we shall get something like Fig. 9.2.

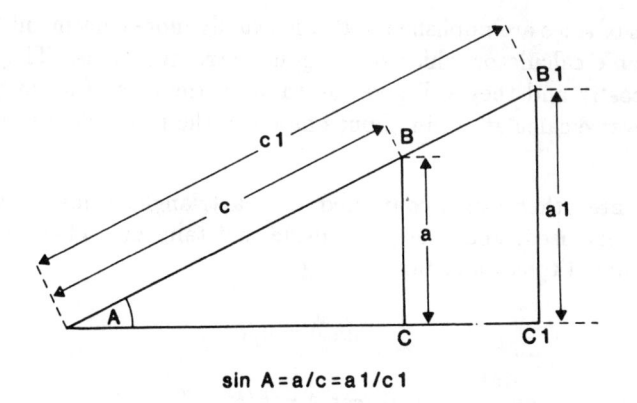

sin A = a/c = a1/c1

Figure 9.2 Right angled triangle, scaling effect

Now two triangles which have got the same angles, but are different sizes are called similar triangles, and their sides are all in the same ratio. If you like you can consider one to be a scale model of the other. Because of this the ratios a/c and a_1/c_1 are equal, so

$$\sin A = \frac{a}{c} = \frac{a_1}{c_1}$$

This means that for the particular angle A, $\sin A$ is always the same no matter what the size of the triangle. If we want we can build up a table of values of a/c or $\sin A$, for different values of A. So if we know the value of a/c we can find the the angle and vice versa.

sin A = a/c   cos A = b/c   tan A = a/b

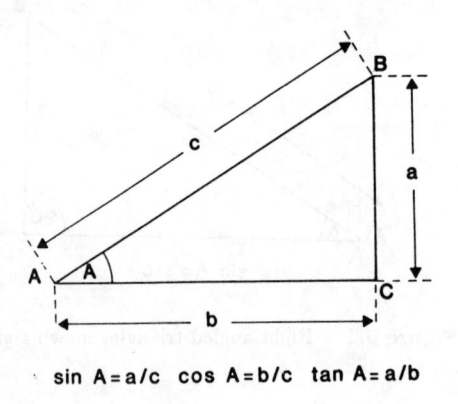

Figure 9.3   Right angled triangle, showing $\sin A$, $\cos A$ and $\tan A$

Tables of sines are published, but it is usually more convenient to use an electronic calculator which has trigonometric functions. They are not very costly and they will give the value of the sine of an angle, or do the reverse calculation, in about one tenth the time taken to look up a table.

There are other ratios connected with a triangle, some of which are commonly used, and these are cosine and tangent. They are usually abbreviated to cos and tan (see Fig. 9.3).

$$\sin A = a/c$$

$$\cos A = b/c$$

$$\tan A = a/b$$

9: Surveying Calculations

It is useful to have some idea of the behaviour of these ratios as the size of the angle varies. This can prevent you making silly mistakes.

When the angle A is small (Fig. 9.4)

$$\sin A \text{ becomes small}$$
$$\tan A \text{ becomes small}$$
$$\cos A \text{ approaches } 1$$

When the angle A is nearly 90° (Fig. 9.4)

$$\sin A \text{ approaches } 1$$
$$\tan A \text{ becomes very large}$$
$$\cos A \text{ becomes small}$$

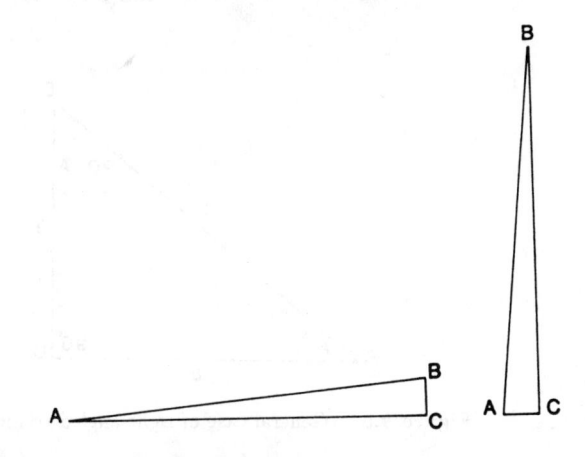

Figure 9.4 Limiting case for $A \to 0°$ and $A \to 90°$

At the limiting values

$$\sin 0° = 0 \quad\quad \sin 90° = 1$$
$$\tan 0° = 0 \quad\quad \tan 90° = \infty*$$
$$\cos 0° = 1 \quad\quad \sin 90° = 0$$

---

* ∞ this is the symbol for infinity (or bigger than you can imagine).

The sum of the angles of a triangle is always equal to 180°, therefore in Fig. 9.5.

$$A + B + C = 180°$$

but since $\quad C = 90°$

it follows that $\quad A + B = 90°$

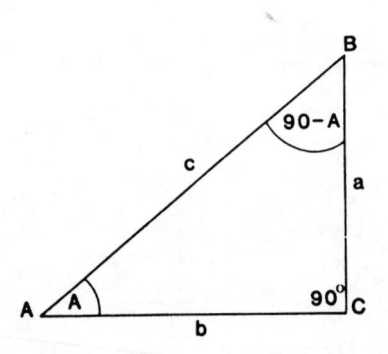

Now $\quad \sin A = a/c$

but $\quad \cos B = a/c$

and $\quad B = 90° - A$

so $\quad \sin A = \cos(90° - A)$

and $\quad \tan A = \dfrac{1}{\tan(90° - A)}$

Figure 9.5 General case of right angled triangle

9.3 Large Angles and Bearings

So far we have looked at angles which are somewhere between 0° and 90°. Now it is time to look at larger angles, and at the same time we might as well introduce a new way of looking at the angles. The mathematician likes to think of angles in the way shown in Fig. 9.6.

He imagines angles getting bigger by rotating one of the lines anti-clockwise. The surveyor, on the other hand, likes to think of angles measured from North in a clockwise direction, as in Fig. 9.7.

If he doesn't know where true north is exactly, he chooses some direction to suit himself and measures his angles with reference to that direction. He then says that the angle from this chosen direction is the bearing of the line.

9: *Surveying Calculations*   81

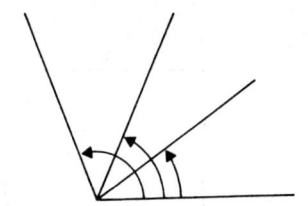

Figure 9.6   Mathematical bearing convention - from x direction

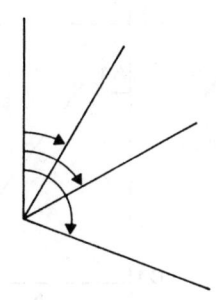

Figure 9.7   Surveying bearing convention - from North

In Figure 9.8

$$\sin A = \frac{BC}{AB} = \frac{a}{c}$$

$$\sin A_1 = \frac{B_1 C_1}{AB_1} = \frac{a}{c}$$

The angle $C_1 AB_1$ is equal to the angle CAB. So we can say that

$$\sin A_1 = \sin(180° - A_1) = \sin A$$

Here we can introduce another useful device which helps us to keep track of our position on a plan. Surveyors agree on a convention that a movement to the North is positive, while movement to the South is negative. Similarly movement to the East is positive and to the West negative.

In going from A to B in Fig. 9.9 we move positively along the $N - S$ or Y axis and negatively along the $E - W$ or X axis. The surveyor very often refers to the $N - S$ direction as *latitude* and $E - W$ direction as *departure*.

The $X - X$ and $Y - Y$ axes divide the area around A into four parts and these are referred to as the quadrants, *N.E.*, *S.E.*, *S.W.* and *N.W.*

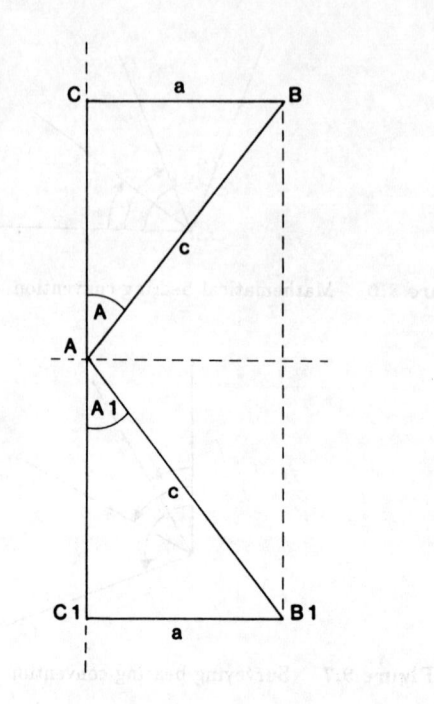

Figure 9.8 Sin of angle between 90° and 180°

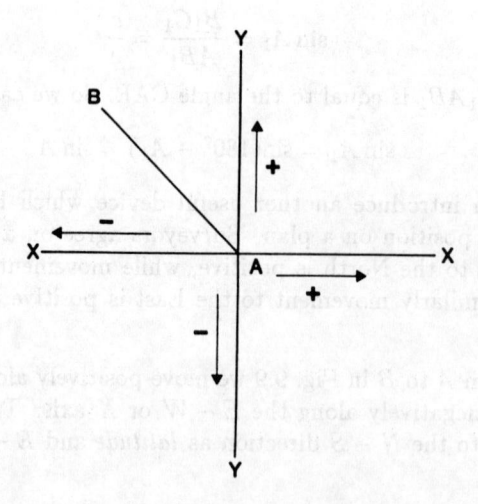

Figure 9.9 Sign convention, latitude and departure

9: Surveying Calculations

We have already seen what happens to the value of the sine of an angle if the bearing of the line lies in the $N.E.$ and $S.E.$ quadrants. Let us now extend the investigation to include the $S.W.$ and $N.W.$ quadrants, as in Fig. 9.10.

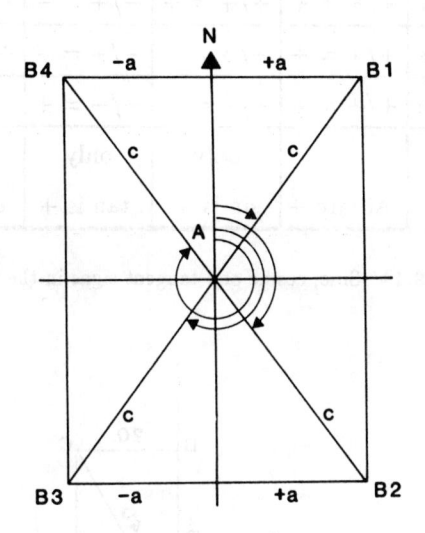

Figure 9.10 Quadrant bearings

In the $S.W.$ and $N.W.$ quadrants the points B_3 and B_4 are just as far away from the $N-S$ axis as B_1 is in the $N.E.$ quadrant. But the signs of both these two distances are negative, while that of B_1 is positive. So in the $S.W.$ and $N.W.$ quadrants $\sin A$ is negative. In the $S.W.$ quadrant $\sin A$ is $-(\sin(A - 180°))$ which is the same as $-\sin A$ (in $N.E.$ quadrant).

Another way of looking at it is to examine the sign of both parts of each ratio, but always taking c as positive. Thus in the $N.E.$ quadrant $\sin A = +a/+c$ which gives a result which is $+$. In the $S.E.$ quadrant we get $+a/+c$ which again gives a $+$ result.

Looking at all three ratios, sin, cos, tan and considering the signs in each quadrant, we can make table 9.1.

Suppose that we have a look at how this works with some numbers. A convenient angle to use is 30°, because $\sin 30° = 0.5$ exactly. If we have a line AC as shown in Fig. 9.11, which is 40m long and the bearing of AC is 30° then we know that

$$BC = AC \sin 30° = 40 \times 0.5 = 20\text{m}$$

Quadrant				
	N.E.	S.E.	S.W.	N.W.
sin	+/+ = +	+/+ = +	−/+ = −	−/+ = −
cos	+/+ = +	−/+ = −	−/+ = −	+/+ = +
tan	+/+ = +	+/− = −	−/− = +	−/+ = −
	All are +	only sin is +	only tan is +	only cos is +

Table 9.1 Sine, cosine and tangent signs in the four quadrants

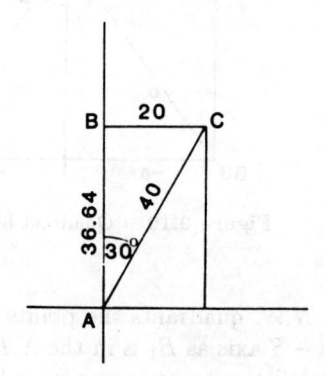

Figure 9.11 Example of 30° bearing

So C lies 20m to the right of the $N-S$ axis through A. We can also find how far C is from the $E-W$ axis through A. In this case we have

$$BA = AC \cos 30° = 40 \times 0.866 = 34.64 \text{m}$$

If, as in Fig. 9.12, the bearing of AC is 150° then these values will come out as

$$BC = AC \sin 150° = 40 \times 0.5 = 20 \text{m}$$
$$AB = AC \cos 150° = 40 \times (-0.866) = -34.64 \text{m}$$

So the sign of AB tells us that we are going in the opposite direction to the first case.

Looking into the case when the bearing is 210°, see Fig. 9.13, we get

$$BC = AC \sin 210° = 40 \times (-0.5) = -20 \text{m}$$

9: Surveying Calculations   85

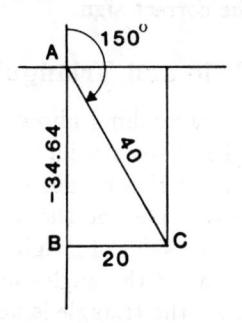

Figure 9.12   Example of 150° bearing

$$AB = AC\cos 210° = 40 \times (-0.866) = -34.64\text{m}$$

These results tell us that we have moved down to the left.

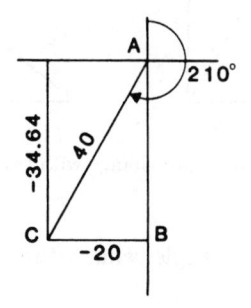

Figure 9.13   Example of 210° bearing

If you are working with a set of tables of sin and cos you will find that very often they only go up to 90°. For larger angles you have to juggle with the value and decide for yourself what the sign should be. An electronic calculator will accept angles up to 360° and it will give the value of the trig. function (sin, tan or cos) and put the correct sign in front. In addition when given the value of the angle and the length of the line it can usually be made to produce the two results as given above without having to feed the same information in twice. The more sophisticated types may be programmed to do the whole operation over and over again, so that you can deal with a lot of values just as quickly as you can punch in the numbers.

It is worth noting that most of the calculators will accept angles greater than 360°. They deduct 360° or 720° and deal with the remainder and still come up with the correct sign.

9.4 The Sine Rule and Triangulation

So now that we know something about trigonometrical functions and how to cope with angles of any size let us look at an important surveying application. Before we do this we need to think in a more general way about sin, cos and tan. We got the value for sin by looking at the ratio of the lengths of the sides of a right-angled triangle and as we saw $\sin 30° = 0.5$. But if one of the angles of a triangle is 30° the value of its sin is still 0.5, even if the triangle is not a right-angled one. You can imagine a right-angled triangle enclosing the real triangle if you like, as in Fig. 9.14.

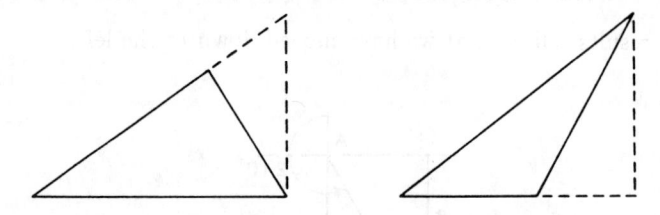

Figure 9.14 Sine does not change with the type of triangle

So the sin, cos and tan of an angle remain the same, no matter where we come across the angle.

If we take any triangle, as in Fig. 9.15, with the angles A, B, and C and sides a, b and c, in which a is opposite A, b is opposite B and c is opposite C we have the following simple rule:-

$$\frac{a}{\sin A} = \frac{b}{\sin B} = \frac{c}{\sin C}$$

Suppose that we measure c and find it to be 137.54m, and we measure the angles A, B and C and they turn out to be

A 57°14′20″

B 67°48′00″

C 54°57′00″

Now $A + B + C$ should be equal to 180°0′0″, but as measured they turn out to be 179°59′20″ which is a good result being only 40″ less than it should be. If the error was large, say more than 2′ we would know

9: Surveying Calculations

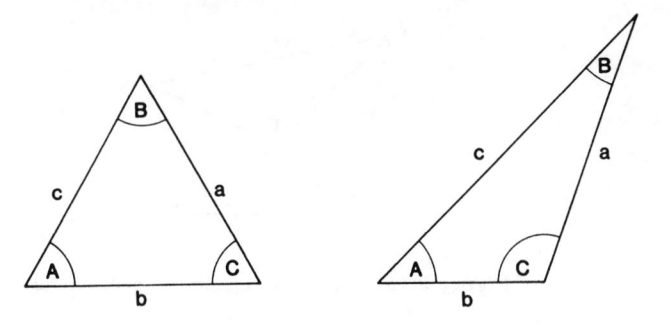

Figure 9.15 The sine rule

that we had been a bit careless and had better remeasure the angles. If it was very large, say 10' or more then we have really made a mess of things. With an error of anything up to 2 minutes then the best thing to do is to just divide the error almost equally among the three angles, which then become

A 57°14'33"

B 67°48'13"

C 54°57'14"

Now we can write

$$\frac{a}{\sin 57°14'33''} = \frac{137.54}{\sin 54°57'14''}$$

therefore

$$a = \sin 57°14'33'' \times \frac{137.54}{\sin 54°57'14''}$$

and likewise

$$b = \sin 67°48'13'' \times \frac{137.54}{\sin 54°57'14''} = 155.550$$

These two calculations for a and b can be done in next to no time with our electronic calculator. We can now think of extending the system to something like that shown in Fig. 9.16.

Unless you are using EDM (Electronic Distance Measuring) equipment, which is very costly, it is far quicker to measure angles by theodolite than it is to measure lengths of sides. All we need to do then is to measure one side and all the angles. Then the Sine Rule opens the way to find the lengths of all the other sides.

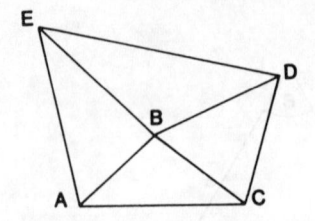

Figure 9.16 Simple triangulated quadrilateral

There is also another great benefit from this procedure. When we measure the angle between two points A and B they may be at very different elevations. The angle that the theodolite records is the true horizontal angle between the two. So if we start with the measurement of the one side as truly horizontal, then all the other lengths that we calculate will be the horizontal values. So we can now put down a number of main survey stations to cover an area that we wish to survey. For example we may have stations A, B, C, D, E, F, G and H, as shown in Fig. 9.17.

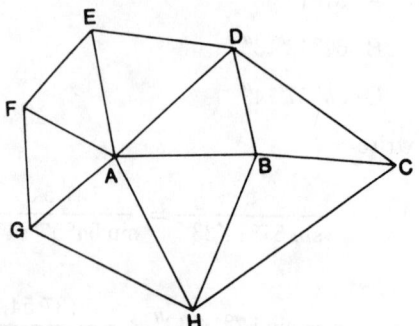

Figure 9.17 Small triangulated network

The position of each station is determined by the detail which will be picked up from it, and by its relation to the other main stations. Sometimes a station may only serve the purpose of providing a connection to other stations. The horizontal length of one line must be measured, so it is useful to look out for a suitable location. Sometimes this may be the only purpose of two of the stations. Now all the angles are measured and each triangle checked for the 180° condition. Minor adjustments may be made to the angles to make each triangle sum to 180°, but if more than minor adjustments are needed then remeasure the doubtful angles. Also the angles at points like A and B must all add up to 360°.

9: Surveying Calculations

In very precise surveying a more rigorous procedure is followed, but for a lot of field work the above will be quite sufficient.

Now by applying the Sine Rule as described before we can find the length of all the sides. A good procedure is to start from the side that was originally measured, calculate right round the system in an orderly fashion and return to the same side that we started from. In Fig. 9.18 the sequence has been shown by the numbers and arrows. So by starting from side 1 and going on to 2, 3, 4 etc. in sequence we shall eventually come back and find a calculated value for 1. Very rarely will it be exactly the same value that we started with, but as long as the two agree closely then we know that all our work checks.

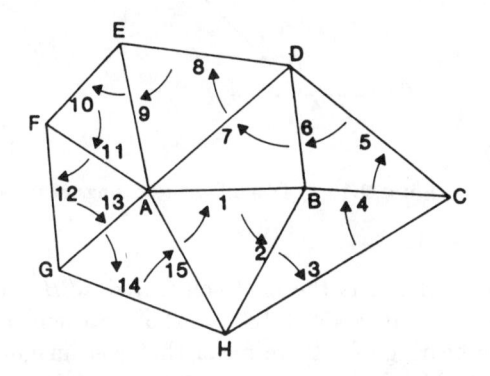

Figure 9.18 Sequence of side calculation

9.5 Plotting Triangulation

Before plotting the detail it will be necessary to draw in the triangulation framework. From our knowledge of the lengths of the sides we can do this using a beam compass if we wish. However this is not always the most convenient way and it may be better to locate the positions of the main stations on a rectangular site grid. This will also be a good method if a tacheometer or theodolite is going to be used for picking up detail.

One of the triangulation sides is chosen and it is given an arbitrary direction. For example if one of the lines is running approximately from South to North then the line is given a bearing of 0°. It may be that a line is running approximately from West to East and in that case the bearing would be 90°. Usually the choice should fall on one of the longer and more important lines. In using bearings it is essential to remember that the direction in which you face is important. The bearing of AB in

Fig. 9.19 means that standing at A and looking towards B the direction is 45° clockwise from North.

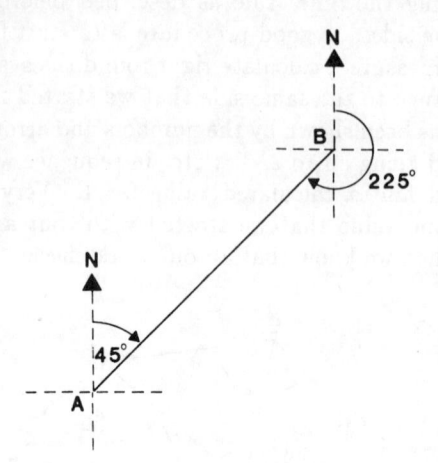

Figure 9.19   Two different bearings of the same line

The bearing of BA is the angle measured at B clockwise from North and in this case it is 45° + 180° = 225°. So remember that if we talk about the bearing of AB we mean that you imagine yourself standing at A and looking towards B. The bearing of BA implies that you are standing at B and looking towards A. The two values will always differ by exactly 180°, although we are talking about the same line all the time.

Starting from the line which has been given an arbitrary direction, and using the adjusted values of the angles which were used for the triangulation calculations, the bearings of all the lines are calculated.

For example, in Fig. 9.20, if the bearing of AB is 90°, add on 57°31′20″ to give the bearing of AC as 147°31′20″. Add on 180° to give the bearing of CA as 327°31′20″. To get the bearing of CD we have to move 67°54′40″ anticlockwise from CA, so this is subtracted to give 259°36′40″.

By working round the system we can return to the original line and calculate either its bearing, or reverse bearing. So once again we have a check on our calculations, and in this case, since we are using adjusted values of the angles the check should be precise.

Now we have a complete list of the lengths of the sides of the triangles and their bearings from one end or the other. From this information we

9: Surveying Calculations

In tabular form

$$\begin{align}
\text{bearing of AB} &= 90° \\
+ &\ 57°31'20'' \\
\text{bearing of AC} &= 147°31'20'' \\
+ &\ 180°00'00'' \\
\text{bearing of CA} &= 327°31'20'' \\
- &\ 67°54'40'' \\
\text{bearing of CD} &= 259°36'40''
\end{align}$$

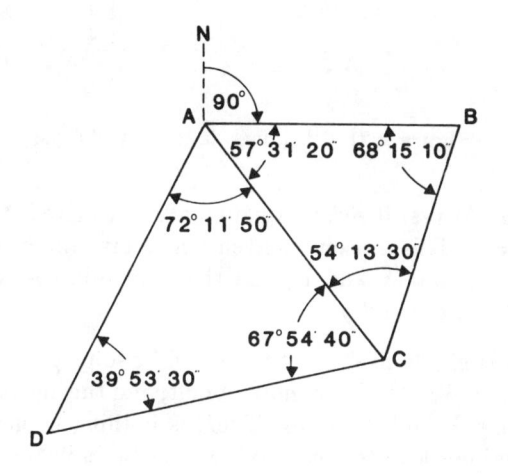

Figure 9.20 Bearings in a small triangulation

can calculate the change in latitude and departure (or x and y) along each line, as explained previously, thus departure (x) = $L\sin\theta$ and latitude (y) = $L\cos\theta$. The angle θ is the bearing and using this we shall get the right signs for x and y.

So for a line AB as in Fig. 9.21, if we use the bearing θ_1 as measured at A, we shall get x and y both positive. What this means is that to travel from A to B we must add on both x and y. If on the other hand we use bearing θ_2 as measured at B, both x and y will come out negative since to move from B to A we have to subtract both x and y. So it is important that we know from which end of the line we are working.

Once we know the x and y values for each line we can start to think about the position of all the points on the site grid. To do this we need to fix the position of one point and then we can work out the position of all the other points in relation to it. A natural choice seems to be one of the points which defined the line to which we first gave the value

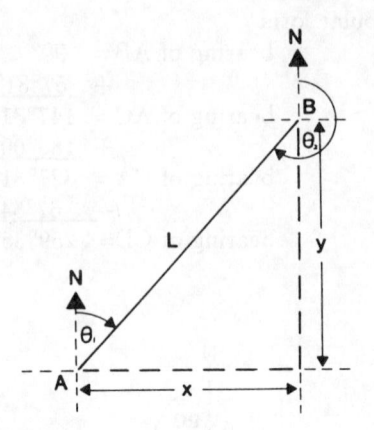

Figure 9.21 Bearings, latitudes and departures

of a bearing. At first it seems equally natural to give it the co-ordinates $X = 0, Y = 0$. It is a useful convention to give the co-ordinate values of points as capital X and Y, and the change in co-ordinates along a line as lower case x and y.

However, it is highly likely that there will be some points below (South) and to the left (West) of our chosen point, and this means that they will have negative X and Y values. There is nothing wrong with this, but it can sometimes lead to confusion and mistakes in the arithmetic. We can get over this difficulty by calling our chosen point not 0, 0 but 400, 800 say. These values are chosen so that all the points on the survey will have positive X and Y values. To get these values we make a rough estimate, tending to err on the large side and then round up to the next 100 above.

In choosing the arbitrary value for our chosen point you should remember that we do not need to include the point 0, 0 on the plot. It may indeed be way, way off the bottom left hand corner of the sheet since we have now arranged our numbers so that there is nothing that we want to plot near it. You may like to think of the grid as plotted on a large sheet of transparent squared paper, with the origin or main intersection $X = 0, Y = 0$, plotted in the middle of it. If we put this over the triangulation plot we can slide it about as we wish, so if we push it over to the left and downwards we make sure that the whole of the survey area is in the upper right (or North East) quadrant.

Suppose that we have arranged for the direction AB in Fig. 9.22 to have a bearing of 90° and point A is at $X = 400, Y = 800$. Let the length of $AB = 174.613$.

9: Surveying Calculations

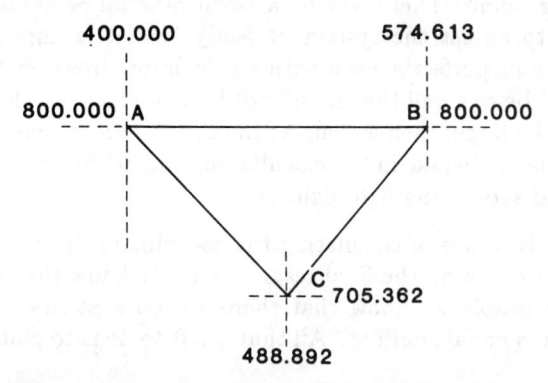

Figure 9.22 Co-ordinates of triangulation points

For the lines BC and CA we have calculated values as follows:-

	x	y
BC	- 85.721	- 94.638
CA	- 88.890	+ 94.641
AB	- 174.613	0.000

Note that for CB, $x = +85.721, y = +94.638$ Now starting at A we have

	x	X	y	Y
A		400.000		800.000
AB	+174.613		0.000	
B		574.613		800.000
BC	- 85.721		-94.638	
C		488.892		705.362
CA	- 88.890		+94.641	
A		400.002		800.003

This shows how to go round all the points and find their X and Y values. Once again we can usually do it in such a way that we can return to our starting point. There will be a small residual error, because it is impossible to adjust the system perfectly, and it is impossible to do the calculations perfectly, even with a calculator. However the residual error should be so small that it will not be visible on the plot. If it does turn out to be large, then as long as the earlier checks were satisfactory there must be a mistake in the calculations and arithmetic. A thorough check should soon bring it to light.

The great advantage of calculating the co-ordinates in this way is that it provides a check on the final result. Once we know that the residual error is acceptable we know that there are no mistakes either in the measurements or calculations. All that is left to do is to plot the points.

9.6 Dealing with Sloping Ground

It is not easy to measure accurately horizontal distances down slopes using tapes and ranging poles. When setting out grids it is important to maintain accuracy on the horizontal distances; this can be readily achieved by the use of simple trigonometrical principles.

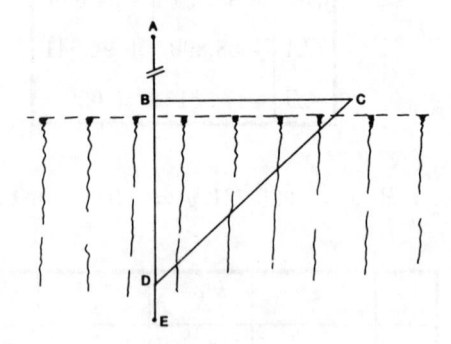

Figure 9.23 Continuing a line down a slope

Suppose that as shown in Fig. 9.23 we wish to continue the line AB down the slope and fix a peg E at some given distance from A. We can fix the point B at the top of the slope and set up a theodolite there. Then we can use the theodolite to put a peg D on the line AB but without trying to put it in at any particular distance. All that we do is ensure that it is down past the bottom of the slope. From B a line is set out at 90° to BD, and at some convenient distance along it a peg is put in at C. Next the theodolite is used to measure the angle at C between B and D.

$$\text{Now} \quad \frac{BD}{BC} = \tan C$$

9: Surveying Calculations

therefore $BD = BC \times \tan C$

Since we know BC and C we can calculate BD. So now we have the distances AB and BD. If they are subtracted from the required length AE what we have left is the length DE. With the theodolite reset at B and zeroed on to D we can run the line forward from D to E with no problems in the linear measurement.

For example suppose we want AE to be 50m. The point B set up at the crest of the slope is found to be 8.715m from A. Point C is set at 40m from B. The angle as measured at C is $43°52'35''$.

$$BD = 40 \times \tan 43°52'35'' = 38.461$$

therefore $AB + BD = 47.176$

and $DE = 50 - 47.176 - 2.824\text{m}$

There is an alternative way to do this which is shown in Fig. 9.24.

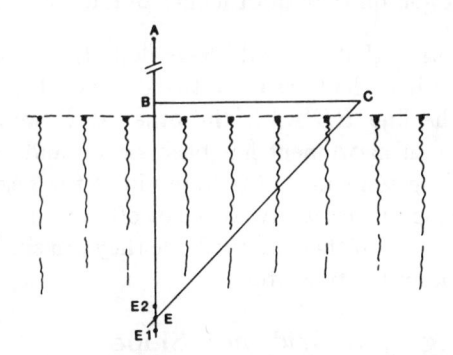

Figure 9.24 Setting out a point down a slope

From A the point B is set up at the crest of the slope as before. Now an approximate postion of E is found, either by doing a rough measurement down the slope or by using any other information available.

From the theodolite position B two points E_1 and E_2 are located on the right line, but lying one on each side of the rough location of E at about 1.5 to 2m apart. Also the point C is fixed as before at some convenient distance from B.

By subtracting AB from the required distance AE we know what the length BE should be.

$$\frac{BE}{BC} = \tan C$$

and so we can find what the angle C should be when we have E in the correct position. The theodolite is set up at C and the instrument zeroed on B. Then the angle C is turned. Now by sighting through the telescope we can fix E between E_1 and E_2. To do this stretch a fine string tightly between E_1 and E_2. A peg can now be sighted in on this line by observing through the telescope. Finally a nail can be put in at the point of intersection on the peg. Note that since we zero on to B, the instrument will swing anti-clockwise to sight E, so the reading will reduce from 360°. In our example (360° − 45°54′20″) = 314°5′40″.

The two methods described for doing this particular task should be studied. We shall come across them again and they are useful methods to employ in other instances as well. Note that in the first case we chose a point almost at random, but connected with the point we wanted to get to, and then having found where our random point was we had only a short measurement to make to put us on to the proper point. In the second method we used two subsidiary pegs to locate part of a line which straddled the desired location, and then we established our point by an intersection on that short length of line.

It is worth noting that we could have done the second method rather more elegantly if we had used two theodolites. One set up at B would sight along the line BE while the other at C would give the intersection. Once an instrument has been set up and aligned a relatively inexperienced person can do the lining in of the peg and nail, so only one skilled person is necessary. It is not often that field sites can boast two theodolites, but if they are available they can simplify the surveying enormously on undulating sites.

9.7 Setting up a Grid on a Slope

It will help in talking about slopes if we borrow two terms from Geology, these are *dip* and *strike*. The dip of the slope as shown in Fig. 9.25, is its maximum angle from the horizontal and the line of dip (sometimes just referred to as dip) is down the line of greatest slope. The strike line is at right-angles to the dip and is horizontal.

If a grid has to be set out on a slope it will usually be most convenient to arrange the grid so that the grid lines are parallel to the dip and strike lines. Then all measurements in the grid direction parallel to the strike will be truly horizontal. Of course those parallel to the other axis will have to be done with care so that they are horizontal. At least 50% of the measurements will not require special precautions. Any other alignment of the grid will mean that all measurements will need special care.

9: Surveying Calculations

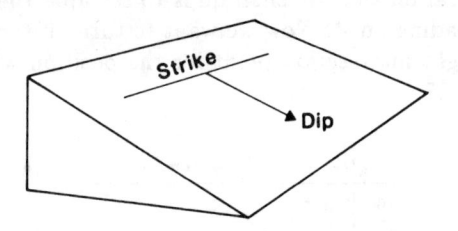

Figure 9.25 Dip and strike

Choose a base line to work from which is parallel to the strike, and if possible position it near the upper part of the area to be covered. From such a position it is possible to look down on to all the points which are to be set out and a clear view, free from obstructions is obtained. Suppose we wish to set out pegs at 10m intervals, then along the chosen base line carefully set out pegs with nails at 10m distances. Since the chosen base line is horizontal this task will be straightforward, but take care because much of the rest of the work depends upon it.

Now we wish to run two lines of pegs down the slope, at right-angles to the base line. Choose two pegs on the base line which are a good distance apart and from which lines running down the slope will not meet any obstructions. Suppose that in a particular case two such pegs are chosen 70m apart. The situation is then something like that shown in Fig. 9.26.

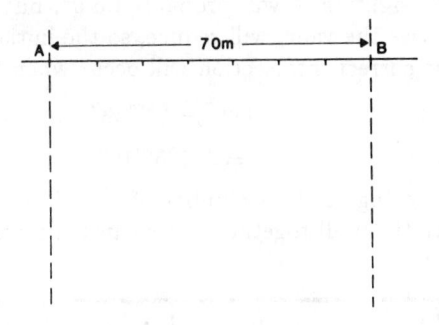

Figure 9.26 Base line with lines at right-angles

Set up a theodolite at A, sight towards B and then turn off 90° in a clockwise direction. Now put in pairs of pegs along the 90° line so that each pair straddles the approximate position of a 10m grid point. This is the technique which was described earlier for fixing a single point.

The next operation that we shall do is to set up a theodolite at B and take a zero reading on A. Now we want to turn off the angle which will give us the right intersection point for the position which is 10m from A. See Fig. 9.27.

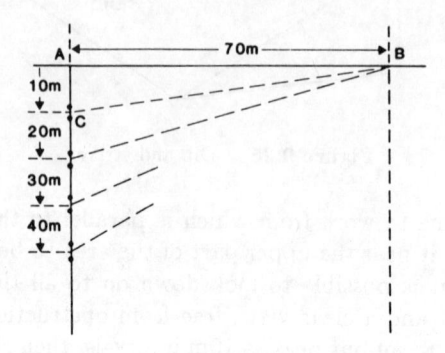

Figure 9.27 Setting points on one line

Now $\dfrac{AC}{AB} = \tan B$

therefore $\tan B = \dfrac{10}{70} = 0.1428571$

and $B = 8°7'48''$

When turning this angle the theodolite will swing anti-clockwise. The reading when sighting A will probably be 0°, but since the movement is anti-clockwise the value will reduce, so the initial reading is taken to be 360°. The correct intersection will occur when the angle reading is

$$360° - 8°7'48''$$
$$= 351°52'12''$$

Since we are going to do a number of such intersections we might as well calculate them all together so we can build up a table.

Int.	tan B	Angle B	Reading
1	10/70=0.1428571	8°7'48''	351°52'12''
2	20/70=0.2857143	15°56'43''	344°3'17''
3	30/70=0.4285714	23°11'55''	336°48'5'
4	40/70=0.5714286	29°44'42''	330°15'18''

9: Surveying Calculations

Notice that the values of tan B go up by equal amounts, but the corresponding angles do not go up by equal amounts, that is why each one has to be calculated separately.

Having set the points we need along the line through A we can do the same procedure for the line through B. In this case the rôles of A and B are interchanged. Now of course the angles as set out at A are the same as calculated above for B, and they are not subtracted from $360°$. The accuracy of the work may be checked by measuring between the two lowest points on the A and B lines. The distance should be within an acceptable limit of error of 70m. In addition the angle between this line and the A or B line may be checked.

As described above only one theodolite is used. If two theodolites are available the work may be done much more quickly. Three people can put in the pegs in a remarkably short time and to a high degree of accuracy.

When using this method it is best not to try to put in too many points down the slope from the base line. Once the angle required gets to about $45°$ the accuracy begins to fall off. However, in this case the working base line may be transferred to one of the strike lines set out lower down the slope, and then the whole sequence of operations repeated.

9.8 Problems with Large Grids

If a very large area is under consideration, problems are likely to arise because of the topography of the site or because of the fragmented nature of the excavations. Before we look at these in detail it will be useful to familiarise ourselves with some of the calculations that may crop up.

Suppose that we are interested in a point Q with co-ordinates as in Fig. 9.28

$$X_Q = 428.13 \qquad Y_Q = 862.74$$

and we are investigating its position with respect to a main grid intersection P with co-ordinates

$$X_P = 410.00 \qquad Y_P = 850.00$$

The length PQ is equal to the square root of the sum of the squares of the differences in co-ordinates (Pythagoras). The calculation is as follows

$$X_Q - X_P = 428.13 - 410.00 = 18.13$$
$$Y_Q - Y_P = 862.74 - 850.00 = 12.74$$

100   *9: Surveying Calculations*

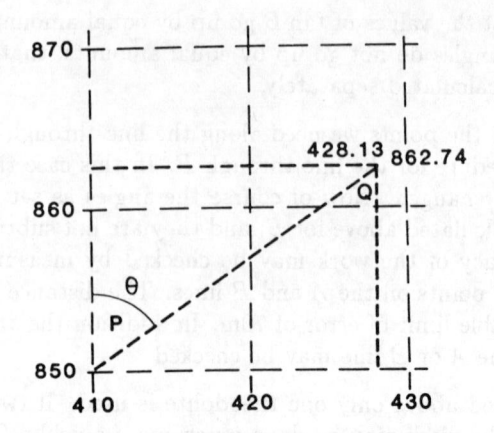

Figure 9.28   Co-ordinates of point in grid

$$18.13^2 = 328.6969 \qquad 12.74^2 = 162.3076$$
$$18.13^2 + 12.74^2 = 491.0045$$
$$\text{Therefore} \quad PQ = \sqrt{491.0045} = 22.159$$

All of which can be done quickly on the electronic calculator. With many of them you won't have to write anything down, the machine will take care of it all.

Also we can find the angle θ, quite easily since

$$\tan \theta = \frac{18.13}{12.74} = 1.4230769$$

$$\text{Therefore} \quad \theta = 54°54'15''$$

and in this case this is the bearing of PQ. Of course the bearing of QP is

$$180° + 54°54'15'' = 234°54'15''$$

Another type of calculation involves three points on the grid. Suppose that these are P, Q and R as illustrated in Fig. 9.29. P and Q are known points (either main grid points or not) and R is a peg whose position we wish to find.

P and Q have the following co-ordinates

$$X_P = 834.88 \qquad Y_P = 217.96$$

$$X_Q = 856.07 \qquad Y_Q = 224.81$$

9: Surveying Calculations

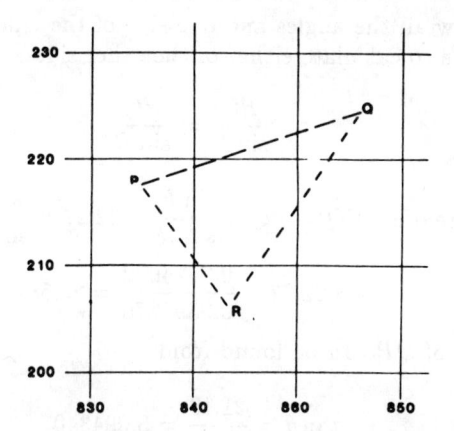

Figure 9.29 Co-ordinates of three points in grid

Assume that the angles at P, Q and R are measured and found to be

$$P \quad 72°14'35''$$
$$Q \quad 37°50'5''$$
$$R \quad \underline{69°53'50''}$$
$$179°58'30''$$

Now adjust P, Q and R to give

$$P \quad 72°15'05''$$
$$Q \quad 37°50'35''$$
$$R \quad \underline{69°54'20''}$$
$$180°00'00''$$

It is not essential to measure all three angles, since from any two values the third can be calculated, but then there is no check on the accuracy of the work.

Now calculate the length PQ as follows

$$X_Q - X_P = 21.19 \quad (X_Q - X_P)^2 = 449.0161$$
$$Y_Q - Y_P = 6.85 \quad (Y_Q - Y_P)^2 = 46.9225$$
$$\text{Therefore} \quad PQ = \sqrt{495.9386} = 22.27$$

We now know all the angles and one side of the triangle so we can use the Sine Rule to calculate either of the other sides.

$$\frac{RG}{\sin P} = \frac{PQ}{\sin R}$$

Therefore $RQ = PQ \times \dfrac{\sin P}{\sin R} = 22.27 \times \dfrac{\sin 72°15'5''}{\sin 69°54'20''}$

$= 22.27 \times \dfrac{0.9524032}{0.9391276} = 22.58$

The bearing of QP can be found from

$$\tan \theta = \frac{21.19}{6.85} = 3.0943407$$

Therefore $\theta = 72°5'8''$

Bearing $QP = 180° + 72°5'8'' = 252°5'8''$

Bearing $QR = 252°5'8'' - Q$

$= 252°5'8'' - 37°50'35''$

$= 214°14'33''$

$X_Q - X_R = QR \times \sin 214°14'33'' = 12.706$

$Y_Q - Y_R = QR \times \cos 214°14'33'' = 18.666$

Now $X_R = X_Q - 12.71 = 843.36$

$Y_R = Y_Q - 18.67 = 206.14$

Sometimes it may happen that P and Q are not intervisible. In this case when set up at P and Q we use some other known point such as S as a zero. We can then calculate the bearing of point S from P say, and also the bearing of Q from P.

Bearing PQ − Bearing PS = Angle SPQ

We measure the angle SPR and then we can find angle QPR since

$$QPR = SPR - SPQ$$

All this seems to be an inordinate amount of calculation, but it must be emphasised that an electronic calculator makes short work of it. Keep an orderly check on paper of what values you have and write out your

9: Surveying Calculations

proposed steps if necessary; then let the calculator do the tedious bits. Now we are in a position to look at some of our problems in detail.

Suppose that we are dealing with a large area and we wish to put down the main intersection points of a grid. We can think in terms of an area about 2000m by 1000m with grid intersections at about 20m intervals. In an area of this size we can expect to have ridges, valleys and other irregularities in the ground which will make it very difficult to run a straight line through from one end to the other. To maintain reasonable accuracy a lot of time will be taken in being very careful measuring up and down the slopes. Also the theodolite will have to be moved forward frequently to maintain visibility as the line goes over crests and down hollows. Of course if the area is almost a uniform plane (and there are areas where this may be the case), then bless your luck and carry on. For the other less tractable sites try the method suggested here.

The general idea is to put in a series of grid points at about 100 - 200m intervals. Then the intermediate ones can be put in more easily between these, and the accuracy of their alignment and distance does not need to be high since the errors will not be propagated right across the site. Each square of 100 - 200m side will act as a check and control for the subsequent operations.

But how are we to put in our control points? The strategy here is to put out a triangulation system over the whole area since this will give a number of accurately fixed positions, and then from these points to establish the control points.

First of all look over the whole site noting possible places for a base line and triangulation stations. This will probably take at least two hours walking round the site. On the first circuit just walk around and get a general impression and on the second one start putting in ranging poles in possible positions. Do not commit yourself too strongly to them at first, give them a critical appraisal. Try to see how you will use them first as triangulation stations and secondly as points to establish the main 100 - 200m grid points. With regard to the latter function remember that once you have put in some of the main grid points they may also be used to fix others.

Once the triangulation network has been settled measure the base line and all the triangulation angles. Then work out the length of sides and decide on the alignment of the grid. Next work out the co-ordinates of the triangulation stations on the grid. This may sound like a lot of work but it is a much quicker and more accurate way than measuring lines over bumpy ground.

By this stage you should know what spacing of main grid control points you wish to use. Suppose that it is 150m. Draw out a plan showing the 150m grid lines and on this plan plot the triangulation points. This does not need to be a large or elaborate plot. It is only intended to help us decide how we are going to set about the job of putting in the control grid intersection points, and what bearings we need to calculate.

Fig. 9.30 shows part of such a plot and how from A and H we could fix the points (150,0) and (150,150). From the example of calculations given earlier we saw how to calculate the necessary bearings. Then with two theodolites, one set up at A and the other at H, we can easily and accurately fix these and other points in that area of the grid. If there is only one theodolite available then we must resort to putting marks each side of the required position, but in this case we shall probably have to use ranging poles rather than pegs and string.

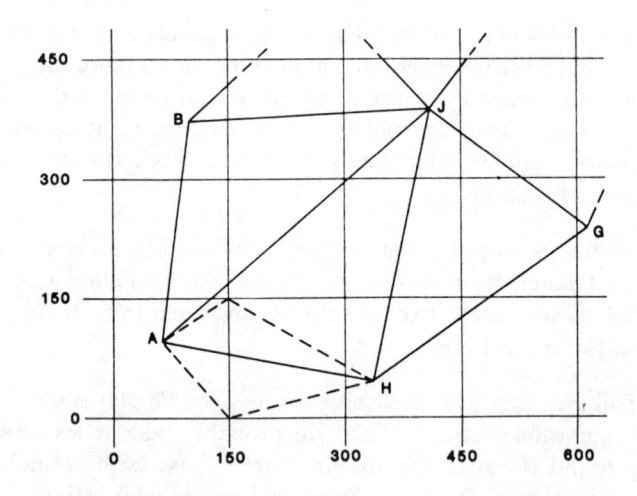

Figure 9.30 Setting out a corner of a large grid

Remember that the main pegs themselves can be brought into play, for example study Fig 9.31. It shows that from two main grid points separated by three main grid intervals it is theoretically possible to get twelve good intersections on each side, giving a total of twenty-four possibilities. The topography will probably intervene to reduce this number considerably. It must be remembered that grid intersection locations are dictated by the alignment of the grid, and so will often fall on points difficult to sight to or from. This is why the triangulation system is so useful, since the selection of station positions to command the best areas possible is at the choice of the surveyor. However, the

9: Surveying Calculations

illustration given shows that a useful strategy is to go for one or two interior lines first. On a small site which is fairly flat it may be possible to do the majority of the site in this way with an absolute minimum of linear measurement. As stated before, once the main control points have been established there should be little difficulty in putting in the intermediate grid positions. There are times when there is no need to set out the whole of the grid. For example there may be an extensive area with several excavation sites which are possibly inter-related. It will be most useful if each excavation grid conforms with the main grid pattern, but there is no need for the grid to cover the intermediate areas.

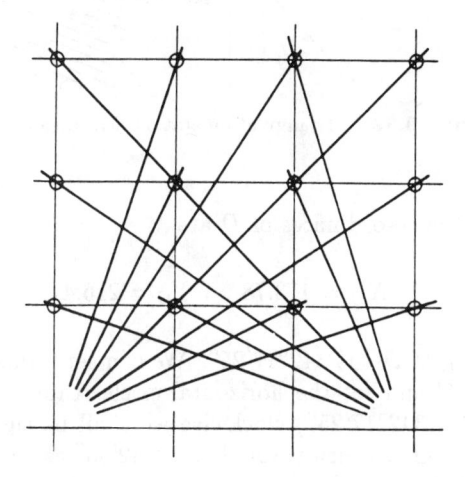

Figure 9.31   Setting out grid points from grid points

To deal with this problem we can start as we did in the previous case by laying out a base line and triangulation system. In choosing the positions of the triangulation stations we should try to put some of them close to the areas of interest, but this should not be allowed to result in badly shaped triangles. Then the base line must be measured and all the angles of the triangulation system measured and adjusted. Following this we can calculate the lengths of all the sides. Now the choice of grid alignment must be made, and in this type of site it is very important that an informed choice is made. It is almost certain that there will have to be compromises between the requirements of the different areas and so careful consideration must be given to the matter. Also any correlation that may be sought between the areas may be helped or hindered by the alignment of the grid. Following this decision the bearings of all the sides and co-ordinates of the stations can be calculated.

Now let us look at the case of a triangulation station which fortunately we have managed to get near to an area of interest, as in Fig. 9.32.

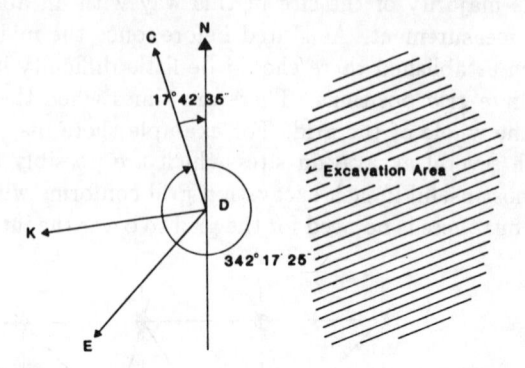

Figure 9.32 Triangulation station near area of interest

Suppose that the co-ordinates of D are

$$X_D = 473.78 \qquad Y_D = 216.44$$

and the bearing of DC is $342°17'25''$. We can set a theodolite up at D, sight towards C and set the horizontal circle reading to zero. Now if we turn $(360° - 342°17'25'')$ clockwise we shall be sighting along grid north. So the angle to turn through is $17°42'35''$ as shown in Fig. 9.32. Since D is 216.44m north, if we measure out $(220 - 216.44) = 3.56$m in the direction that the theodolite is set we shall have fixed a point on the 220 N line. Now add $180°$ on to the theodolite reading to give $180° + 17°42'35''$. The theodolite is now pointing along grid south. By measuring $(10 - 3.56) = 6.44$ and 16.44m in this direction we find the 210 and 200m N lines. Now we can transfer to the latter point, sight station D and turn off $90°$. Then if we measure $(480 - 473.78) = 6.22$m we find the 480 E line. We are also at 200 N, so this point is (480,200). The pegs that we have in are as in Fig. 9.33.

Now by using two steel tapes, or with the theodolite and steel tape, other pegs can be put in at convenient locations around this particular area. In some cases it may not be possible to arrange for one of the main triangulation stations to be immediately adjacent to an area. Where this happens we can put in a subsidiary station close to the area that we wish to service and link it to the triangulation system. The method of linking in this subsidiary station depends upon the local topography, but generally it will be best to fix it by using two of the main triangulation stations to form a triangle with the subsidiary point, as in Fig. 9.34.

9: Surveying Calculations   107

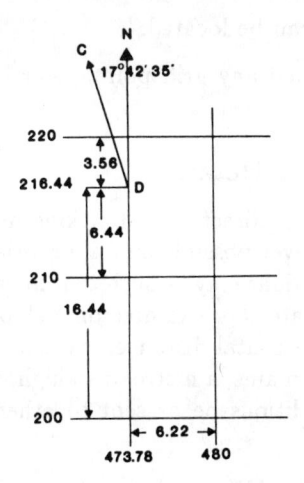

Figure 9.33   Main grid pegs

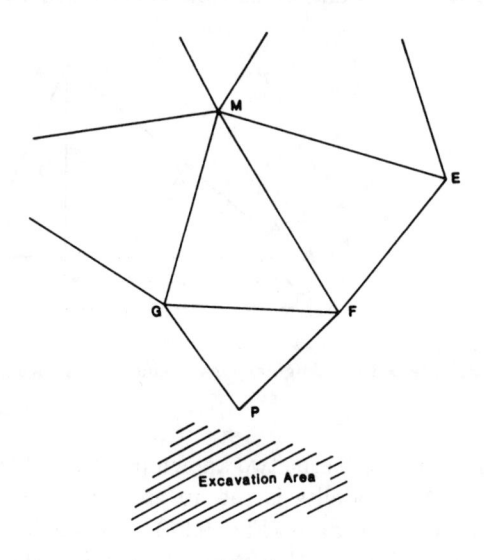

Figure 9.34   Triangle with subsidiary point

Thus the point P forms a triangle with F and G. If all the angles are measured then the position of P can be found in relation to F or G by applying the same principles that have been used in the main triangulation.

Alternatively, if for example F and P are not intervisible, then the angle FGP should be measured and the line GP carefully measured. From this information P can be located.

Once P has been-fixed any grid points near by can be set out as described before.

9.9 Surveying of Detail

The simplest and most direct way of picking up detail is by linear measurement, and wherever possible and appropriate you should do it this way. However, conditions may be such that linear measurement becomes difficult and inaccurate. For example on a sloping site it becomes difficult to get a true horizontal distance, even in only a moderate wind. If there are standing remains, a metre or so high, the same problem arises. Should the two conditions be present together then life becomes very difficult.

An accurate method, which can be used under such adverse conditions, is to make angular measurements from two known points. Before going into the practical details let us look at the calculations involved.

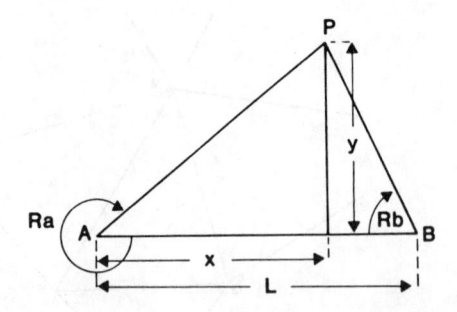

Figure 9.35 Angular measurement from two known points

Suppose A and B to be the two known points, and the distance between them is L. A theodolite is set up at A and set to zero on B, and then turned to point at P (Fig. 9.35). The angle recorded will be the clockwise angle AB to AP. Let us call this R_A (the reading at A). The angle A in the triangle will be equal to $(360° - R_A)$. A similar procedure at B, but with zero setting on A will give a reading R_B, which will be equal to the angle B in the triangle.

In the triangle, angle A + angle B + angle P = 180°.

$$\text{Therefore} \quad (360° - R_A) + R_B + P = 180°$$

$$\text{Therefore} \quad P = R_A - R_B - 180°$$

9: Surveying Calculations

By the Sine Rule
$$\frac{AP}{\sin B} = \frac{AB}{\sin P}$$

Therefore $AP = AB \times \dfrac{\sin B}{\sin P}$

$$= \frac{L \times \sin R_B}{\sin(R_A - R_B - 180)}$$

$$x = AP \cos R_A \quad \text{and} \quad y = -AP \sin R_A$$

The values of x and y may be found quite quickly using an electronic calculator, and the formulæ are in such a form that they will give the correct signs as well. A is taken to be the origin of the axes and wherever P lies in relation to AB, x and y will have the appropriate signs. For example if P is below AB then y will have a minus sign.

If a lot of points have been taken then it will pay to put the results on to a computer. Depending upon the computer installation it may be possible to get the points plotted on a graph-plotter. If this can be done it is very well worthwhile, as then all that remains to be done is to join up the points with the appropriate lines and then make a fair copy.

Now let us look at some of the practical points connected with the method. If possible choose the two points of observation relatively high up, as then they will command a good view of the points to be surveyed. Someone will have to hold a ranging pole at all the points in sequence keeping it as near to the vertical as possible. The theodolite should be sighted on the lowest point of the pole which is visible. From a high observation position it will frequently be possible to sight the very bottom of the pole.

The assistant who is going to hold the ranging pole should be escorted around the points which should be marked by chalk, arrows, or other suitable means. The points should be surveyed in sequences of about 15 - 20 otherwise the assistant might get confused. A sketch showing the points will be a help to the assistant and if one is provided the number of points in a sequence can probably be increased.

The angles from one station are observed and then the whole operation repeated from the other station. If two theodolites are available then the angles may be read simultaneously from both stations, and the work proceeds twice as fast. In such a case it pays to ensure that the two observers do not get out of step with the numbers of the points. This can be done by getting the assistant to give a signal at every tenth point as a check for the observers.

If there are some well defined features on the site which can easily be measured then they may prove useful checks on the work. Similarly local detail may be measured and noted on a small supplementary sketch.

After a little experience a team of three people should be able to record about 40 points in an hour.

As always in surveying work the field records should be made in a neat and orderly fashion. If particular attention is paid to this it may pay dividends when processing the results by computer. It may be possible to hand in the field readings to be entered without further copying and to receive in return a list of the co-ordinates and a drawing of the points all plotted in their correct positions.

9.10 Traverse Calculations

A traverse consists of a series of straight lines which are connected end to end. If it returns to the starting point it is called a *closed traverse*, but if not then it is known as an *open traverse*. The main elements of a traverse are the lengths of the lines and the angles at each intersection point between the two lines which meet there.

To plot a traverse it is convenient to calculate the co-ordinates of each point on some appropriate grid system. If the traverse stands alone as a surveying framework, then the grid may be aligned with respect to one of the sides, preferably one of the longer and more important ones. If the traverse is part of a larger framework then at least one station (usually the first) must be common to both and angles must be taken there to relate the bearings on the traverse to those of the whole system. In either case it will be possible to find the bearing of any line of the traverse with respect to the grid being used.

Let the bearing of a line of a traverse, measured from grid north, be θ, and let the length of the line be L. Then in moving from one end of the line to the other the change in departure (or x) will be $L\sin\theta$, and the change in latitude (or y) will be $L\cos\theta$. The sign to be given in each case will depend upon the direction of movement and the quadrant in which θ lies. If these quantities are calculated for each line of a traverse and they are successively added with due regard to sign (latitudes and departures being kept separate), then the values recorded after each pair of additions will be co-ordinates of the intersection points of the traverse.

Fig 9.36 shows an open traverse D, a, b, c, d which starts from station D on a triangulation system. The bearing of the line DE as determined from the triangulation is $332°41'20''$. The first stage is to calculate the

9: Surveying Calculations

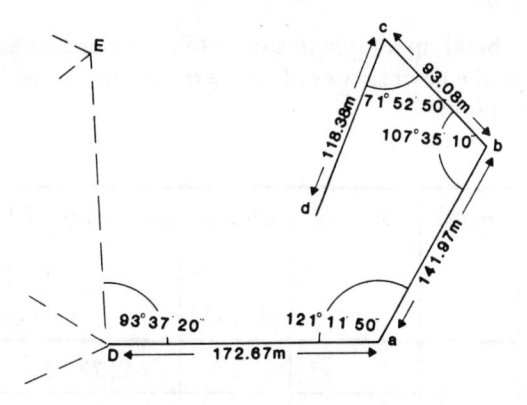

Figure 9.36 Traverse calculations

bearings of all the lines in the traverse. Remember that to reverse the bearing of a line 180° may be added, or subtracted, whichever is most convenient.

Bearing of DE	332° 41′ 20″
Angle Eda to be added	93° 37′ 20″
	426° 18′ 40″
Since result exceeds 360° deduct	360° 00′ 00″
Bearing of Da	66° 18′ 40″*
To reverse bearing of Da add	180° 00′ 00″
Bearing of aD	246° 18′ 40″
Angle Dab to be added	121° 11′ 50″
	367° 30′ 30″
Since result exceeds 360° deduct	360° 00′ 00″
Bearing of ab	7° 30′ 30″*
To reverse bearing of ab add	180° 00′ 00″
Bearing of ba	187° 30′ 30″
Angle abc to be added	107° 35′ 10″
Bearing of bc	295° 05′ 40″*
To reverse bearing of bc deduct	180° 00′ 00″
Bearing of cb	115° 05′ 40″
Angle bcd to be added	71° 52′ 50″
Bearing of cd	186° 58′ 30″*

The required bearings are marked thus *

Note that if it was a closed traverse we would get a check on the result since we should return to the original bearing, within the limit of allowable error.

Now we can build up a table to calculate the co-ordinates of the points. Assume that the co-ordinates of D are departure (x) 642.73m and latitude (y) 394.06 m

Point	Line	Length	Bearing	Change of dep. ($L \sin \theta$)	Departure	Change of lat. ($L \cos \theta$)	Latitude
D					642.73		394.06
	Da	172.67	66°18'40"	158.12		69.37	
a					800.85		463.43
	ab	141.97	7°30'30"	18.55		140.75	
b					819.40		604.18
	bc	93.08	295°05'40"	-84.29		39.48	
c					735.11		643.66
	cd	118.38	186°58'30"	-14.38		-117.50	
d					720.73		526.16

Again it should be noted that if this was a closed traverse the calculations should lead back to the starting point, and the values of the co-ordinates return to the original values within a reasonable degree of error.

Chapter 10

EDM, Total Stations and Computers

10.1 Introduction

The use of electronic equipment in Surveying is spreading rapidly. Only a short time ago both computer installations and field equipment were large and required very skilled operatives. Now personal computers are common and field equipment is relatively light and easy to operate. Anyone capable of using a video recorder or a Hi-Fi system will soon master the basic operating procedures of a Total Station or EDM equipment. Once you have done this the equipment can be used to carry out the surveying operations as described earlier in this book. The basic principles to follow do not change; they are independent of the equipment used. The use of electronic equipment will make the work easier, faster and more accurate. Computers may be used at all stages of a survey from recording field data to drawing the final plan.

10.2 Electronic Distance Measurement

This is the name given to various systems that apply electronics to the measurement of distance. They all use some kind of beam which is sent out from a transmitter and then returned from a reflector back to the transmitter. A wave form is superimposed on the beam as it leaves on its outward journey. When it arrives back a comparison is made between the waves going out and those returning. From this the time of travel is found, and when applied to the speed of the beam the distance is known.

The system just described is affected by variations in atmospheric temperature and pressure which means that a correction has to be applied. This may be done either by adjusting the instrument, which will then give the correct reading, or by applying a factor to the value recorded. In both cases it is usual for the instrument to have as an accessory a table, a graph or a circular slide rule which may be used to find the correction for given atmospheric conditions. For the most accurate work it is necessary to have reliable values for the temperature and pressure. The correction is expressed in parts per million and is usually small so that for a lot of work the exact values of temperature and pressure are not critical. The sensitivity of the correction to variations in the temperature and pressure may be explored by trying a range of values in the particular device supplied with the instrument. This knowledge will then be of use when deciding what to do in a given practical situation.

The transmitter is usually supported above the telescope of a theodolite on a mounting which can be adapted to most makes of instrument (Fig.10.1). The reflecting prism may be carried at the top of a special rod or fastened on the top of a tripod. With a single prism the maximum range is usually between 1 and 2km depending upon the particular model and the weather conditions.

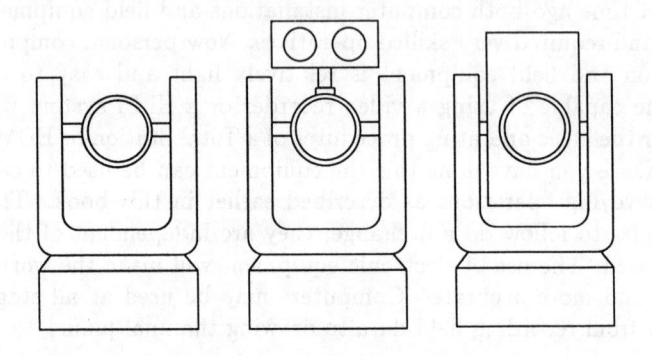

Figure 10.1   Theodolite, theodolite with EDM, and total station

The distance which is recorded is the sight length between the transmitter and the reflector. The instrument takes several readings of the distance, usually about eight, and then displays the mean value. Unless the sight is horizontal it will be necessary to record the vertical angle in order to find the horizontal and vertical distances. If the theodolite being used is an electronic type there may be provision for a link to the EDM and an automatic read-out of the horizontal and vertical distances.

10.3 Total Station

In this type of instrument an electronic theodolite and an EDM are built as one combined unit in a single housing, sometimes sharing the same optical system (Fig.10.1). Distances, vertical angles and horizontal angles are all presented in a digital form. The simplest models give the horizontal and vertical angles along with the sight length. Some have a small auxiliary calculator for finding the true horizontal and vertical distances. The sight length is automatically fed into the calculator and when the operator enters the vertical angle the required results

10: EDM, Total Stations and Computers

are displayed. More complex and expensive instruments give the true horizontal and vertical distances along with other information which will be useful in carrying out certain surveying operations.

All total stations may be connected to some form of automatic recording device, which means that the operator does not need to write anything down. The information from the data recorder may be transmitted directly into a computer, either on or off site, for the results to be processed. A portable computer may be used to receive the data directly from a total station. This allows the surveyor to do some preliminary processing of the data. Any problems are then revealed at a very early stage and can be dealt with immediately.

10.4 Setting Up

The method for setting up a Total Station is the same as that described for a theodolite in Chapter 7, but if you are using a theodolite fitted with EDM it may be necessary to vary the procedure a little. The EDM and its associated battery may make it more difficult to carry out the centring and levelling routines. If this is the case then it is better to set up the theodolite by itself but only approximately centred and levelled. The EDM and its battery may now be fitted and in doing so it does not matter if the setting of the instrument is disturbed. The final adjustments can then be done quite quickly.

10.5 Taking Readings

As with all electronic equipment it is necessary to press keys to operate the instruments and display the results. There is no standard pattern for this; it varies from maker to maker. However they are all very similar since the surveyor requires the same information irrespective of the make of instrument that is being used. To illustrate what you may expect to see we will have a look at two different types. Real instruments were chosen as models, but for convenience of reproduction certain changes were made to layout and labelling. One of the instruments is a very basic type and the other is a more complex total station.

The most basic EDM will only measure the sight distance and this is the type which might be fitted to an optical theodolite. The controls are as shown in Fig. 10.2.

There is a rotary switch which can be set to OFF, MEASURE, TRACK, BATTERY and REF SIGNAL. Turning this switch to MEASURE puts it in the standard measurment mode. To start the measurement process the button marked MEASURE is pressed lightly. If you stab at the button it is likely that the instrument will tilt forward. The needle on

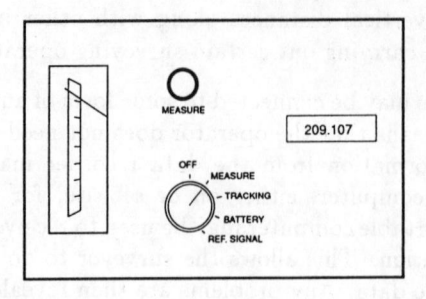

Figure 10.2   Simple EDM controls

the meter at the left hand side flickers as the instrument goes through its sequence of reading and finally the result is displayed in the window at the right hand side. If the rotary swich is set to TRACK and the button is pressed the instrument displays each individual reading it takes; it does not calculate the average of a series. This facility is useful if the equipment is being used for setting out as the prism can be moved backwards and forwards quickly until the right distance is found. The BATTERY setting allows you to check the state of charge of the battery. This is shown by the position of the needle on the meter at the left hand side. In a similar fashion the REF SIGNAL setting shows whether a strong signal is being received back from the reflecting prism.

A typical key pad of a total station is shown in Fig.10.3.

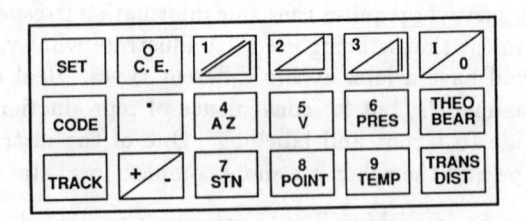

Figure 10.3   Total station key pad

There are quite a number of keys and in most cases the individual keys have a dual function. Some instruments use one key as a switch to change the functions of the other keys; others call the alternative

10: EDM, Total Stations and Computers

functions by using one press or two in quick succession. The purpose of the keys is often indicated by pictograms. Their meanings are not always immediately obvious, but a study of the manual will usually make the logic clear. In most cases abbreviations rather than pictograms are used here.

One key acts as an on/off switch, labelled I/O in our example. The key marked THEO will change the system to theodolite working only, and vice versa. It is almost universal practice to have a picture of a right angled triangle on three of the keys. The sides of the triangle represent the horizontal, vertical and sight distances to the target and on each key a different side is emphasised. If the key with the emphasised horizontal side is pressed then the length of that side is displayed, and similarly for the others. The keys labelled AZ and V are used to call for the values of the horizontal and vertical angles respectively. The abbreviation AZ stands for Azimuth, or horizontal direction.

As we saw in Section 7.6 we need to be able to set the initial horizontal circle reading to some chosen value. The key marked SET used in conjunction with the numeric keys allows us to do this. If the data is to be transferred to an automatic recorder then the key labelled TRANS is used. Such data will need to be identified during the subsequent processing and this may be done by using the CODE key followed by the appropriate numbers. The numeric keys are also used with those marked TEMP and PRES to enter the temperature and pressure for the atmospheric correction. The TRACK key provides the same facility that we met on the basic EDM described above. The correction of an entry is done by means of the CE key.

For some purposes it is useful to be able to enter the co-ordinates of the instrument station and this may be done by pressing the STN key followed by the numeric values. If the co-ordinates of some other point are known then they may be entered in a similar fashion by using the POINT key. The +/- key allows positive or negative values to be used. The keys marked BEAR and DIST will give the value of the bearing and distance from the station to the point being considered. With this facility the zero of the horizontal circle may be set to grid zero and then angles recorded will be grid bearings.

Some total stations have a larger repertoire of functions than those given above and these are useful to the professional surveyor. As long as you have a grasp of the basic functions as described in our model instruments you will be able to make good use of this type of equipment.

10.6 Accuracy of EDM

EDM forms an integral part of all Total Stations so the following discussion should be taken as being equally applicable to them.

Manufacturers of this type of equipment publish values for standard deviations as a constant plus a small fraction of the measured length stated as parts per million. Typical values range from 3mm+3ppm to 5mm+5ppm depending upon the quality and price of the instrument. Hence for measurements of a distance of 100m the standard deviations vary from 3mm+0.3mm to 5mm+0.5mm while for a distance of 1000m the figures are 3mm+3mm and 5mm+5mm respectively. Assuming that we are using one of the cheaper instruments and that we are not as skillful as the manufacturers' surveyors we might expect to achieve accuracies of about 8mm at a distance of 100m and 15mm at 1000m. These accuracies will be good enough to meet the requirement that errors are not large enough to show on the plot. The largest scale that is likely to be used is 1/50 and such a scale will only be used for surveys of a small area with correspondingly small distances to measure. At this scale errors of less than about 10 to 12mm will not show on the plan. A study of the expression for the standard deviation shows that the possibility of errors showing on the plot decreases as the extent of the survey increases.

In addition to being accurate EDM measures distances very quickly. The time to take a set of readings and display the mean is usually of the order of five to eight seconds. To achieve the same accuracy when measuring a length of about 100m with a 30m tape requires care and takes time even if the ground is level. When the ground is not level the problems are compounded. With EDM the type of ground is of no consequence so long as a sight line is possible.

10.7 Basic Co-ordinates

The axis of an EDM which is fitted to a theodolite lies a little above the optical axis of the telescope. To compensate for this the reflecting prism is offset by an equal amount above the sighting target. This ensures that the distance as measured by the EDM is equal to the sight distance. This measurement and the values of the vertical and horizontal angles are required in order to calculate the position of the observed point. A Total Station takes the same fundamental measurements and operates on them internally to display whatever is requested, as for example the horizontal and vertical distances.

The relationship between sight distance, vertical angle, horizontal distance and vertical distance is shown in Fig.10.4. If the sight distance is

10: EDM, Total Stations and Computers

D, and the vertical angle is ϕ, then the vertical and horizontal distances are given by

$$V = D \sin \phi \quad \text{and} \quad H = D \cos \phi$$

These formulae for V and H form part of the internal functions available with Total Stations.

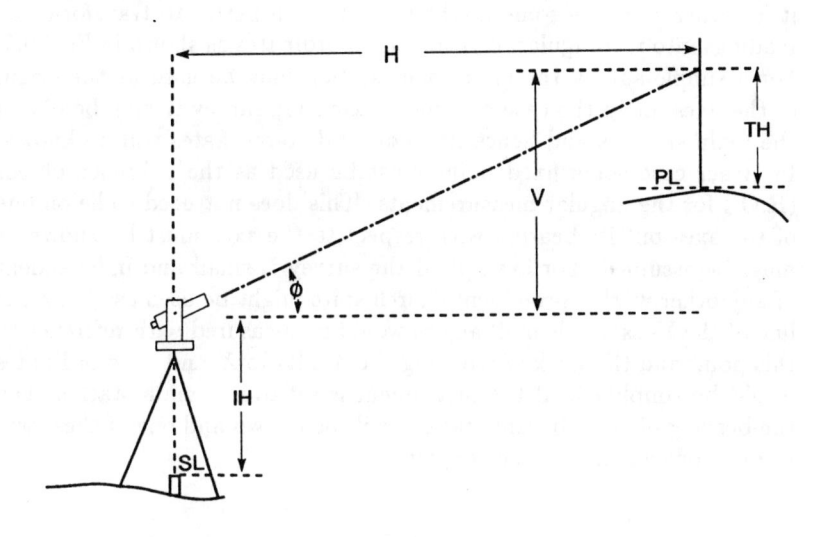

Figure 10.4   Vertical elevation of EDM and target

If the instrument is set at a height IH, above a station of reduced level SL, and the height of the target on its pole is TH, then the reduced level of the observed point PL is given by

$$PL = SL + IH \pm V - TH$$

This formula applies both to Total Stations and theodolites fitted with EDM. It is very similar to that given in section 8.3 for dealing with the fundamental measurements in tacheometry. However in the present case the result is more accurate because of the method used to measure the distance.

It should be noted that if the target is set at the same height as the instrument then these two quantities cancel out in the formula for the determination of the reduced level of the point. It is good practice to do this, but undulations of the ground or dense vegetation may sometimes make it impossible. In such cases make suitable records in your field notes.

The horizontal angle θ, and the horizontal distance H, are used to locate the observed point on the plan or horizontal plane. Used in this way they are similar to the mathematician's polar co-ordinates, but the mathematician uses anti-clockwise angles. It would be possible to plot the results using a protractor and scale rule. Large protractors are available but even if they are used the plot will not be very accurate and it is rather a cumbersome method. It is much better to transform the readings into rectangular or X and Y co-ordinates as shown in Fig.10.5. For a simple survey the instrument station may be used as the origin of the axes, or in the case of a more extensive survey it may be one of the main stations and hence its main grid co-ordinates will be known. In either case some fixed point must be used as the reference object (R.O.) for the angular measurements. This does not need to lie on one of the axes but its bearing with respect to the axes must be known or must be assumed. For example if the survey is small and independent of any other work a prominent church spire might be used as giving the line of the Y-axis. Then all angles would be measured with reference to this point and the work of reducing the results to X and Y co-ordinates would be simplified. If the instrument is set over a main station then the bearing of the adjacent stations will be known and one of these will form a convenient reference object.

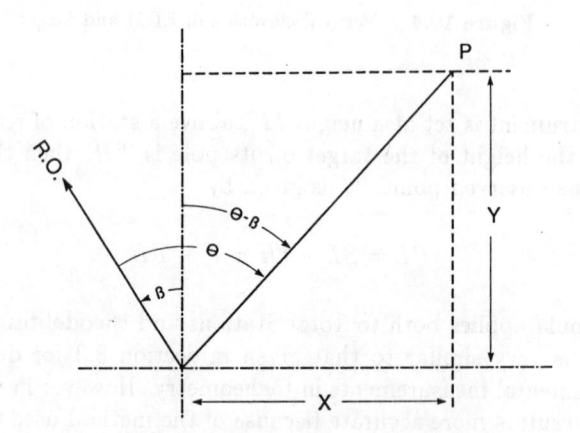

Figure 10.5 Co-ordinate relationships

In Fig.10.5 the angle between the R.O. and the point P is θ and the angle between the R.O. and the Y-axis is β. Hence the bearing of the line OP on the grid is θ-β and this is used to calculate the co-ordinates

10: EDM, Total Stations and Computers

of P,
$$X = H\sin(\theta - \beta) \qquad Y = H\cos(\theta - \beta)$$

For a small local survey these calculations can be done quickly on an electronic calculator.

10.8 Applications of EDM and Total Stations

Electronic equipment is particularly useful in the early stages of a surveying project when dealing with the measurement of the framework. Even if the framework is a conventional triangulation system relying on angular measurement the speed with which EDM can provide a reliable value for the base line-length makes its use worth while. But it can do more than this as we shall see later on.

Electronic instruments may be used for the survey of physical features, topography, contours and other detail. With closely spaced points of detail the assistant can get to the next point before the person on the instrument has written down the last set of readings and time is wasted. If automatic recording is used writing is eliminated and along with it the possibility of mis-booking.

Setting out major grids is made easier if electronic instruments are used and they may be used to set out trenches, but caution will be required when the distances are small. We shall look at this point in more detail later.

10.9 Picking Up Detail

With traditional equipment the survey of detail is in nearly all cases based upon the idea of a line. Frameworks are arranged so that there is always a line near enough to the detail points for the linear measurements to be made quickly and easily. This leads to a strong linear approach both in the field work and the recording.

In many ways EDM is like a fast, accurate tacheometer. This means that it may be used to survey detail not only in those cases where a tacheometer would be suitable, but also where a tacheometer would not be sufficiently accurate. The area of interest lies all round the station, roughly circular in shape with the instrument at the centre. This change from a linear to an area method has an effect upon the style of booking.

A good sketch plan should be made of the area around the station showing the detail required. The points observed can be marked on this sketch plan so that later when they are plotted they may be interpreted correctly. The values of the angles and distances found as each point is observed should be entered in a properly drawn up table and the

122   10: EDM, Total Stations and Computers

identifying code for each point carefully recorded. The code to use will depend upon the way in which the results will be plotted. If it is going to be a manual plot then the best system is to have a consecutive series of numbers so that one number can only refer to one point on the site. If the results are to be drawn by computer then the code will have to tell the computer what type of feature it has to draw and it is likely that this will be laid down in the drawing package used.

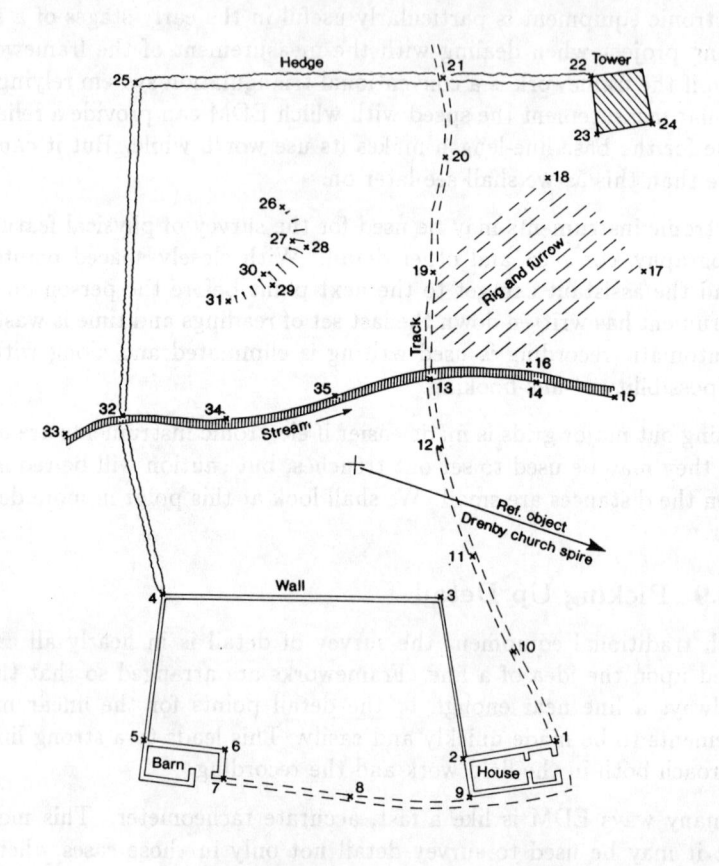

Figure 10.6   Sketch plan

An example of a sketch plan is shown in Fig.10.6. Making a good sketch plan comes with practice and drawing them makes you study the area carefully. As you get more experience you will find that they give you the ability to assess the surveying problems presented by an area and the best way to deal with them.

As compared with a tacheometer, or any other detailing method EDM

10: EDM, Total Stations and Computers

has a much greater range. This has an important effect upon the framework used for a survey. It means that the stations may be more widely spaced. The distance between stations will now usually be limited more by the effect of topography on clear sight lines, and the need to maintain good communication between the surveyor and the assistant.

The thought of using such expensive, high technology equipment for surveying detail may appear surprising, but its accuracy, speed and ease of reading make it very suitable for such work. As long as it is not required elsewhere for dealing with framework measurement and the like then it should be used on whatever work needs doing. The high cost is a reason for employing it to the full, not a reason for keeping it on a shelf. If it is only used for dealing with the main framework it will spend most of its time in its box and be an expensive luxury. For most archaeological work, rather than buying a very sophisticated instrument it is better to get a basic type. This will be perfectly adequate for all requirements and the money saved can be used to buy additional prism sets. Their use will reduce the time required for measurements on main frameworks, and will show even bigger savings for detail work.

10.10 Setting Out

As mentioned at the beginning of this book archæologists, and other field workers, frequently make use of rectangular grids to control their work. Setting out such a grid over a large area can be difficult and suggestions as to how traditional equipment may be used to do it are given in Chapter 9. When electronic equipment is available then it becomes much easier as can be seen from Fig.10.7. If the bearings of the first three lines are calculated, then from the symmetry of the figure about the main 45° diagonal and using simple arithmetic the bearings of the other lines may be found. Similarly it is only necessary to calculate the lengths of the lines up to and including the same diagonal. The other three quadrants are just repetitions of the first. As you become familiar with thinking in terms of radius and angle other ways of developing the full potential of the equipment will become apparent.

Setting out trenches and other short distance work should be approached with caution. It was stated earlier that for some equipment the value of the standard deviation is 5mm + 5ppm. If you are dealing with a length of about 30m an error of 5mm is not very good when the ground is relatively flat: you should be able to do better than that with a steel tape. It is a good idea to test how well you can do with your equipment by setting out one or two short trial lengths using EDM and then comparing them with tape measurements. Of course if the ground slopes steeply or is very rough it will probably be quicker and more

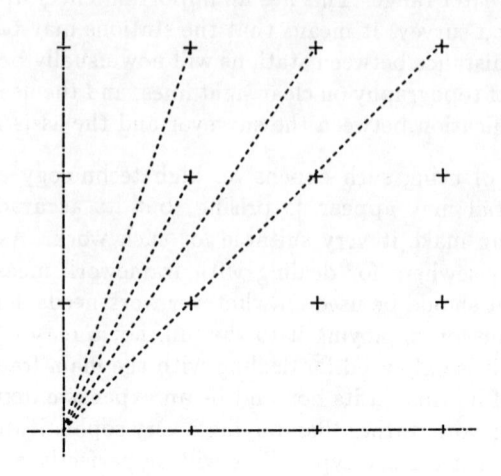

Figure 10.7   Setting out a grid

accurate to use EDM. The problem that you will face then is whether to risk expensive equipment in an area where your own footing is not very sure!

10.11   Linear and Angular Measurement of Frameworks

Although as we saw earlier the time to take a measurement with EDM is remarkably short, it is important to think about the time taken to carry the reflecting prism to the distant station or move it from one station to the next. When this is taken into account the speed of linear measurement is still much greater than by conventional means, but it loses a little of its apparent gain on angular measurement.

The framework adopted for the site and the way in which the work is carried out will have a bearing on the relative advantages of linear to angular measurement. The stations may be indicated by fixing ranging poles vertically on top of them, supporting the poles with stays fastened to tent pegs, or making them visible in some other way. In this case a single person with a theodolite can observe the angles to the adjacent points with no further assistance, and it does not take long to sweep through a round of angles. They can then work their way round the system taking angles at every station. If they are measuring the distances with EDM then they will have to wait while the prism is carried from point to point. Should a second prism be available it will speed up the work but now there will be three people engaged on the operation. Also remember that if a person reading angles is given an assistant to

10: EDM, Total Stations and Computers

book the work the speed will very nearly double.

10.12 Frameworks

In Chapter 3 we talked about common forms of survey frameworks such as linear triangulation, angular triangulation and traversing. At that stage linear triangulation was considered as measurement of the sides of triangles by means of chains or tapes. The use of EDM brings a different perspective on these systems, because now the speed and accuracy of linear measurement approaches that of angular measurement. For this reason when EDM is used to measure the sides of a system of triangles we shall speak of trilateration, and call measuring the angles triangulation.

A survey of an area about 3km by 2km would be large by archaeological standards, but it could be covered by a simple trilateration system using very basic EDM equipment. Such equipment would be quite capable of giving the standard of accuracy likely to be required. Station positions would be dictated by the need to obtain good viewing points and lines for picking up detail, rather than any technical problems connected with the equipment. This means that although EDM will be a great help in establishing the framework the detail survey methods used may prevent its full potential being realised. We shall look at this in more detail later.

Unfortunately there is no quick simple check of the trilateration as there is in triangulation. In the latter case if the sum of the angles in each triangle is sufficiently close to 180° then you can be reasonably sure that there are no major problems. With trilateration it will take either a detailed calculation or a careful plot to prove the soundness of the work.

Although we have spoken of triangulation and trilateration as though they are two completely separate operations, there is no reason why they should not be combined. The same instrument setting will serve for either, and reading the angle as well as the distance will not take very much more time. As pointed out above, it takes time to move the prism and while this is being done the angle can be read. The protection against errors that such a procedure gives will far outweigh the little extra time taken. In fact the ability to measure distances as accurately and quickly as angles allows us to get away from the traditional concept of a system of triangles added on to one another side to side. We can think more in terms of measuring lines and bearings to give position. The work may be checked by using other lines and bearings to the same points or by looking at any triangular figures in the system.

126   10: EDM, Total Stations and Computers

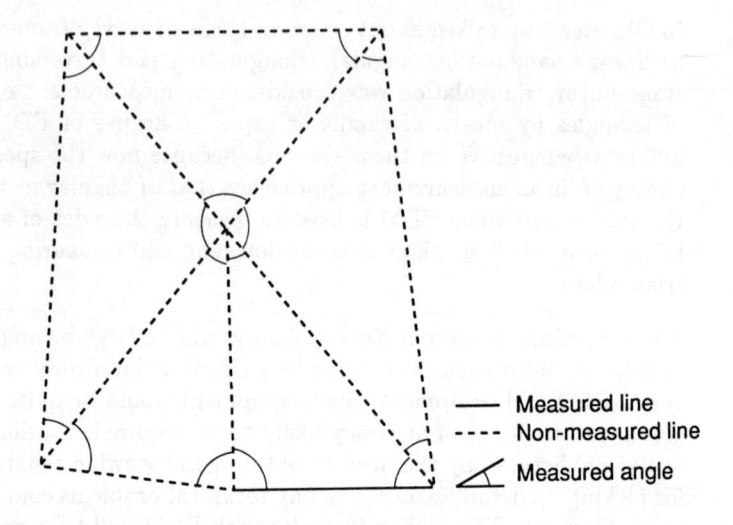

—— Measured line
---- Non-measured line
∠ Measured angle

Figure 10.8   Triangulation using polygon with central station

A simple polygon with central station, as shown in Fig. 10.8 will often be a useful framework to adopt for a conventional triangulation survey of a small area. All the angles are measured and the length of one side found either directly or indirectly. If EDM is available the same framework may be used but the pattern of measurement will be different as shown in Fig. 10.9. From the central station the angles and distances to all the other stations can be measured. This will enable their positions to be found but there will not be a check on the work. If the instrument is moved to one or two of the outer stations, then from these positions readings can be taken to the remaining points to get enough information to verify their accuracy. Which points are chosen will depend upon the intervisibility between the stations.

One of the main problems in the use of the traverse as a framework has been the accurate measurement of the lengths of the lines by conventional means, especially over rough or undulating ground. The use of EDM immediately gets over this difficulty and greatly reduces the time taken to establish this type of framework. To control the driving of the Channel Tunnel two such traverses were run within the bore of the tunnel. They zig-zagged from side to side, crossing and re-crossing

10: EDM, Total Stations and Computers

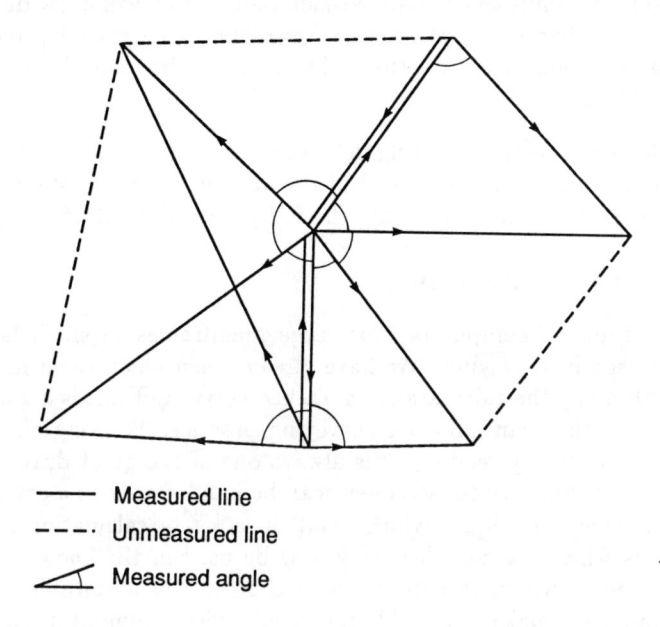

— Measured line
--- Unmeasured line
∠ Measured angle

Figure 10.9 Trilateration using polygon with central station

one another, and providing a continuous check on the position of the working face at any time.

10.13 Communication

Since EDM makes it possible to measure long distances easily and accurately, there is a natural tendency to pick up detail at greater distances from the stations than would be usual with traditional methods. This may lead to problems in communicating with the assistant who is going round with the prism.

When using the more traditional detailing methods the assistant is usually not very far away. In such cases information may be exchanged either by shouting, or by hand signals. Two people who are used to working together will usually have a complex system of hand signals and will know what the other person is doing. These methods become more difficult as the distances increase. The person at the instrument can use the telescope to see the assistant's signals, but the assistant cannot see the signals meant for him. This problem arises at a distance of about 400-500 metres depending upon the eyesight of the assistant. One solution is to give him a pair of binoculars. Another possibility is to use small coloured flags; yellow usually shows up best. Probably the best solution is to use walkie-talkie radio which will allow direct speech. Basic rechargeable sets are now available quite cheaply, and they will quickly repay the investment in terms of time saved and frustration avoided.

There will be time wasting delays and great frustration if the person at the instrument loses control of the actions of the assistant. This is an important factor in realising the full potential of EDM equipment.

10.14 Computers

All types of computers from large mainframes to small laptops may be used in surveying. We have already mentioned them in connection with doing the calculations in Tacheometry, and this is a good example of how they can influence surveying practice. Working out the results of tacheometry readings was always one of the great drawbacks of the method, but now the drudgery can be handed over to a computer. Any surveying techniques which result in a lot of calculations, or calculations which are complex, may now be used in the knowledge that the computer will take care of these chores. With suitable programmes computers make it possible for people whose computational ability or mathematical knowledge is limited to extend the range of their surveying.

Data may be recorded electronically and passed on in that form to a computer. Writing the results in a field book and transcribing them to working sheets is the slowest part of surveying. It is also the most fruitful source of errors: of all mistakes in surveying about ninety percent happen at this stage. Electronic transcription will show a great saving in time and improve the accuracy of data transfer. If an electronic recording system is used it must be capable of producing a hard copy of the field readings, and the memory should not be erased until either such a copy has been made or there is clear evidence that the data is correct.

10.15 Computer Software

This is the field which is changing most rapidly and so it is impossible to give specific recommendations which will remain valid for very long. The main questions to consider are as follows. What surveying problems do you wish to put on computer? What computing equipment is available to you? Do you wish to have the computer plot the survey? What graphical equipment is available to you? For example, if you only have a small PC then it is no use buying a sophisticated commercial surveying

10: EDM, Total Stations and Computers

package; it just will not fit on to your machine. But if somehow you can get a few simple Basic programmes they will run both on your machine and much bigger ones. They will not date and will give good sevice long after a commercial package has been superceded. If you have a big machine and can afford a good commercial package then get one: it will give you far more features and much greater flexibility.

There are quite a number of packages available and deciding which to buy may be difficult. It is a good idea to draw up a list of what you consider to be essential items and a list of things which are not so important but which might be useful.

The essential requirement is the ability to process the data from EDM or total station and give the x,y,z co-ordinates from the station point, and to do the same for tacheometer readings. The next stage on from that is to give the result in terms of the overall grid co-ordinates. It is also essential to be able to work out the details of a traverse and a simple triangulation system. Most packages will offer these two facilities and will probably incorporate some method of adjusting the results. If you know enough about surveying to differentiate between packages on the basis of the methods of adjustment used then you do not need my advice. Those of more limited surveying ability do not need to worry, the results will usually be indistinguishable on the final plot. A useful feature to have is the ability to deal with points defined by their offsets from a straight line, as in basic field surveying. Many packages will also include solutions to other surveying problems as for example intersection, resection or the three-point problem. If these come along with the package that you get then treat them as a bonus. Look them up in a standard book on surveying. You will usually find that the field work is easy and the computer takes care of the difficult bit, thus widening your surveying repertoire with the minimum of effort.

There are software packages which will take the raw field data, process it and produce a finished plot. The process is not completely automatic; there has to be some human intervention in the system. Errors may creep in to the data and instructions are needed concerning the plot, annotations required, style of printing, feature differentiation and the like. In spite of this these systems do save a tremendous amount of work and give a remarkable flexibility to the handling of the plots. They are very expensive but the prices are tending to come down.

Some integrated packages have a data sorting facility which is very useful. If, for example, several walls which are roughly parallel are being surveyed they may be coded as W1, W2, W3 etc. A few points may be taken along each wall and then as the work progresses more points are taken farther along. Prior to plotting the programme sorts through all the data and assembles it into all items of like code so that they will be plotted sequentially. Another feature which may be found is the ability to assign different 'levels' to various parts of the plot. This is very useful for archaeological work as such levels can correspond with different phases. These may then be called and shown either separately or combined with other phases.

Most packages will also draw contours or three dimensional views if required. If contouring is offered the system usually requires the level values to be supplied at the node points of a rectangular grid. In some cases the option of random levels is given. This latter alternative needs to be treated with caution. It is much easier to programme for contours using data on a rectangular grid. When random levels are supplied the programme interpolates the levels at the nodes of a rectangular grid and then carries on to work out the contours. This means that unless you have a reasonably constant density of random points over the whole site the contour plot will not be very good.

10.16 Conclusion

The use of electronic equipment in surveying will go on expanding and the instruments will become even more sophisticated. Even beginners to the subject should gain as much experience as possible in this particular area of the work. The effort in acquiring the skill and knowledge will be well repaid in by the savings in work and time.

Equipment check list

The following is a list of items that you may need, but the choice from it will depend upon the particular job in hand. If you are going to work a long way from your main centre then the rule should be 'When in doubt, take it'. Also remember that if you are in a remote spot damage to a theodolite or breaking the only steel tape can be disastrous. In such cases take along a spare and consider it a good insurance.

Field

Total Station & tripod
Theodolite & tripod
Ranging poles
Arrows
Steel tape
Plane table & tripod
Abney level or clinometer
Plumb-bob
2-3m flexible rule
Board with clips for sheets
Pegs
Hand hammer
Greasy chalk
25mm nails
Soft pencils
Thin string
Plastic bags to cover instruments
Walkie-talkie radio sets

EDM
Level & tripod
Levelling staff
Chain
Plastic tape
Prismatic compass
Cross-staff or optical square
15cm boat shaped level
Field books or booking sheets
Whistle
Large hammer (7lb)
Stout knife
White chalk
150mm nails
Erasers
Stout cord

Office

Drawing board
Set squares
Drawing instruments
Drafting film
Drawing pins
Pencils
Masking tape

T-square
Protractor
Board clips
Erasers
Paper clips
Electronic calculator
Sellotape

Further Reading

- W H Irvine, *Surveying for Construction*, 2nd ed. 1980. London: McGraw Hill. *Comments: Basic Surveying.*

- A Bannister and S Raymond, *Surveying*, 5th ed. 1985. London: Pitman. *Comments: A good standard text book written for the professional engineer and surveyor. Although parts of it may seem advanced it will give the answer to most problems that are likely to arise.*

- J C Pugh, *Surveying for Field Scientists*, 1975. London: Methuen. *Comments: A useful book to refer to for advanced methods. Do not be put off by some of the complexities; look for the bits that you are interested in. Some things are treated in a cumbersome fashion and very advanced topics are introduced. However the clarity of the diagrams brings out the points discussed.*

- C Taylor, *Fieldwork in Medieval Archaeology*, 1974. London: Batsford. *Comments: Good on surveying earthworks, particularly using cross-staff.*

- J Uren and W F Price, *Surveying for Engineers*, 1978. London: Macmillan. *Comments: More advanced treatment of instruments and methods.*

- W S Whyte and R E Paul, *Basic Metric Surveying*, 3rd ed. 1985. London: Newnes-Butterworth. *Comments: Good on chain surveying, small areas and buildings, levelling and types of theodolites.*

- R J P Wilson, *Land Surveying*, 3rd ed. 1985. London: Macdonald & Evans. *Comments: For work on basic plane surveying. Useful further material in other areas.*

- M A R Cooper, *Modern Theodolites and Levels*, 2nd ed. 1982. London: Granada. *Comments: A comprehensive description of theodolites, levels and their adjustment. If you really want to know how they work this is the book to consult, but there is a large amount technical detail.*

- A H A Hogg, *Surveying for Archaeologists and Other Field Workers*, 1980. London: Croom Helm. *Comments: Very good on chain surveying. Very good general comments on careful booking, checking measurements and constant vigilance for sources of error. The attempt to provide guidance over a wide area of surveying has not been uniformly successful: some parts need treating with care. Cutting pegs from branches of nearby trees could be disastrous. Taping up the tilting screw on a quick set level will lead to a great deal of trouble unless you are very experienced, and in that case you probably would not do it. Do not be downcast by the times given for setting up instruments; such perfection is rare indeed.*

Subject Index

Abney level, 49
accuracy, 3
accuracy of EDM, 118
alidade, 31
angles:
    horizontal, 61, 63, 65
    large, 80
    vertical, 61, 65, 66
angular triangulation, 20
archaeological surveying, 1
arrows, 8
automatic level, 40, 42
axes, 7

backsight, 46
basic EDM, 115
basic levelling, 43
bearing, 90, 91
bearing convention:
    mathematical, 81
    surveying, 81
bearings, 80
    quadrant, 83
bench-marks, 43
    temporary, 43
boning rods, 50
book, field, 45
booking and reduction, 45
bubble, plate, 63

calculations:
    surveying, 76
    tacheometric, 69
    traverse, 110
centring, 62
chains, 15
chasing, contour, 55
clinometer, 50
close contouring, 58
closed traverse, 21
collimation, 45
communication, 127
compass, prismatic, 27
computer processing, 74
computer software, 128
computers, 113, 128
connecting to Ordnance Survey, 28

constant, multiplying, 68
contour chasing, 55
contour interpolation – grids, 56
contour interpolation – spot heights, 56
contouring, close, 58
contours, 53
control points, 103
convention:
    mathematical bearing, 81
    surveying bearing, 81
cos, 77, 78
cross-staff, 27

datum, 42
    Newlyn, 42
    section, 49
departure, 81, 82, 91
detail, 21, 28
    surveying of, 108
dip, 96
Distance Measurement, Electronic, 113
drawings, 30
dumpy-type level, 39

EDM, 113, 121
    accuracy of, 118
    basic, 115
earthworks, 52
Electronic Distance Measurement, 113
electronic theodolite, 60, 61, 64

fall, rise and, 45
false origin, 16
field book, 45
foresight, 46
fork, plumbing, 31
frameworks, 18, 125
    measurement of, 124

grads, 76
grid, 91, 92
grid on a slope, 96
grid, national, 17
grids, 94
    large, 99
    rectangular, 6, 123
ground, sloping, 94

hachures, 52
horizontal angles, 61, 63, 65
horizontal planes, 39

imperial measurement, 4
index compensator, vertical circle, 66
intersection method, 32, 33

key pad, Total Station, 116

large angles, 80
large grids, 99
latitude, 81, 82, 91
level:
    Abney, 49
    automatic, 40, 42
    dumpy-type, 39
    mean sea, 42
    quickset, 40
    spirit, 31
levelling, 38, 63
    basic, 43
levels, 39
line plotting, radial, 36
linear triangulation, 19
Liverpool, Old Dock Sill, 42

mathematical bearing convention, 81
mean sea level, 42
measurement of frameworks, 124
Measurement, Electronic Distance, 113
measurement, imperial, 4
measuring on steep slopes, 14
method:
    intersection, 32, 33
micrometer:
    optical, 60, 65
multiplying constant, 68

national grid, 17
Newlyn datum, 42

object, reference, 72
Old Dock Sill at Liverpool, 42
open traverse, 21
optical micrometer, 60, 65
optical plummet, 61, 62
optical square, 26
optical theodolite, 60, 61, 63, 64
Ordnance Survey, connecting to, 28

origin, 7
    false, 16

picking up of detail, 121
plane table, 31
planes, horizontal, 39
plate bubble, 63
plotting, 29, 73
plotting triangulation, 89
plotting, radial line, 36
plumb-bob, 31, 61
plumbing fork, 31
plumbing rod, 62
plummet:
    optical, 61, 62
points:
    control, 103
    reference, 16
poles, ranging, 8
prismatic compass, 27
processing, computer, 74

quadrant bearings, 83
quickset level, 40

radial line plotting, 36
radians, 76
ranging poles, 8
reading of angle scales, 64
rectangular grids, 6, 123
reduction, booking and, 45
reference object, 72
reference points, 16
rise and fall, 45
rod, plumbing, 62
rods, boning, 50

scale, 4
sea level, mean, 42
section datum, 49
setting out of grid, 8
setting out of right-angle, 10
setting up, 62, 63, 115
setting up of dumpy, 40
setting up of quickset, 41
sin, 77, 78
Sine Rule, 86, 87, 89, 109
sloping ground, 94
software, computer, 128
spirit level, 31

square, optical, 26
stadia wires, 31
station, subsidiary, 106
Station, Total, 113, 114, 121
steep slopes, measuring on, 14
strike, 96
subsidiary station, 106
surveying bearing convention, 81
surveying calculations, 76
surveying of detail, 108
surveying, archaeological, 1
system, triangulation, 71, 103, 105

table, plane, 31
tacheometric calculations, 69
tacheometry, 66, 67
tan, 77, 78
temporary bench-marks, 43
theodolite, 8, 10, 60, 87
    electronic, 60, 61, 64

optical, 60, 61, 63, 64
vernier, 60
Total Station, 113,114,121
Total Station key pad, 116
traverse, 21
traverse calculations, 110
traverse:
    closed, 21
    open, 21
triangulation, 86,125
triangulation system, 71,103,105
triangulation:
    angular, 20
    linear, 19
    plotting, 89
trilateration, 125

vernier theodolite, 60
vertical angles, 61, 65, 66
vertical circle index compensator, 66